Guide to
JCT Design and Build
Contract 2016

RIBA ## Publishing

Sarah Lupton

Guide to JCT Design and Build Contract 2016

© Sarah Lupton, 2017

Published by RIBA Publishing, part of RIBA Enterprises Ltd,
The Old Post Office, St Nicholas Street, Newcastle upon Tyne, NE1 1RH

ISBN 978 1 85946 641 4, 978 1 85946 809 8 (PDF)

The right of Sarah Lupton to be identified as the Author of this Work has been asserted in accordance with the Copyright, Designs and Patents Act 1988 sections 77 and 78.

All rights reserved. No part of this publication may be reproduced, stored in a retrieval system, or transmitted, in any form or by any means, electronic, mechanical, photocopying, recording or otherwise, without prior permission of the copyright owner.

British Library Cataloguing-in-Publication Data
A catalogue record for this book is available from the British Library.

Commissioning Editor: Fay Gibbons
Project Editor: Alasdair Deas
Designed and typeset by Academic + Technical, Bristol, UK
Printed and bound by Page Bros, Norwich, UK
Cover design: Kneath Associates
Cover image: Shutterstock: www.shutterstock.com

While every effort has been made to check the accuracy and quality of the information given in this publication, neither the Author nor the Publisher accept any responsibility for the subsequent use of this information, for any errors or omissions that it may contain, or for any misunderstandings arising from it.

www.ribaenterprises.com

Foreword

Design and build procurement routes, in which the employer engages a single contractor to prepare and/or complete the design and construct the works, now account for a significant percentage of all construction contracts in the UK, and increasingly are the preferred method of procurement. The desire of many clients to secure single-point responsibility is compelling, but the extent to which this is achieved using design and build needs to be understood. This Guide will assist with this point and with many other practical and legal issues surrounding the JCT Design and Build Contract (DB16). This contract adopts the familiar logical layout and clear style, while reflecting current working practices, new legislation and new case law. This Guide covers these changes and all the subsequent revisions to the 2011 edition.

Written in straightforward language, Sarah Lupton's Guide to DB16 takes its lead from her excellent Guide to DB11, to provide a clear, comprehensive and thoroughly up-to-date analysis of the form of contract. The contract's provisions, procedures and supplementary conditions are spelled out authoritatively and are organised by theme, and the Guide points out the important new changes in the contract and illustrates the practical effects of the wording with concise and helpful case summaries. The hard-pressed practitioner familiar with the form will be pleased to see the useful indexes, and will doubtless come to depend on being able to dip quickly into the Guide for specific help during the course of a project.

As well as being an indispensable guide for practitioners, I would also thoroughly recommend it to both architecture and other construction students on the threshold of undertaking their professional examinations. Sarah Lupton's rare combination of being a legally trained architect who also runs the MA in Professional Studies at Cardiff University makes this Guide the ideal student companion.

Neil Gower, Solicitor
Chief Executive, The Joint Contracts Tribunal
December 2016

About the author

Professor Sarah Lupton MA, DipArch, LLM, RIBA, CArb is a partner in Lupton Stellakis and directs the MA in Professional Studies at the Welsh School of Architecture. She is dual qualified as an architect and as a lawyer. She lectures widely on subjects relating to construction law, and is the author of many books including this series on JCT contracts, the *Guide to the RIBA Domestic and Concise Building Contracts*, *Which Contract?* and the 5th edition of *Cornes and Lupton's Design Liability in the Construction Industry*. She contributes regularly to the International Construction Law Review and acts as an arbitrator, adjudicator and expert witness in construction disputes. Sarah is also chair of the CIC's Liability Panel and the CIC Liability Champion.

Contents

	Foreword	iii
	About the author	iv
	Contents	v
	About this Guide	ix
1	**Introduction to design and build procurement**	**1**
	The architect's role	4
	Novation	4
	Consultant switch	4
	Some general principles of design liability	7
2	**About DB16**	**11**
	Key features	12
	Deciding on DB16	14
	Comparison with SBC16	15
	Changes in the 2016 edition	15
3	**Documents**	**19**
	Employer's requirements	20
	Contractor's proposals	22
	Contract sum analysis	22
	BIM and other protocols	23
	Health and safety documents	23
	Bonds	24
	Sub-contract documents	24
	Domestic sub-contracts	24
	Use of documents	25
	Interpretation, definitions	25
	Priority of contract documents	25
	Discrepancies and errors	26
	Employer's requirements and contractor's proposals	28
	Divergences from statutory requirements	28
	Custody and control of documents	29
	Performance bonds and guarantees	29
	Assignment and third party rights	29
	Assignment	29
	Third party rights/warranties	30
	Procedure with respect to third party rights and warranties	32

4	**Obligations of the contractor**	**33**
	The design obligation	36
	Standards and quality	39
	Obligations in respect of quality of sub-contracted work	40
	Compliance with statute	41
	Health and safety legislation	42
5	**Possession and completion**	**45**
	Possession by the contractor	45
	Progress	46
	Completion	47
	Pre-agreed adjustment	49
	Extensions of time	49
	Principle	49
	Procedure	50
	Assessment	53
	Partial possession	55
	Use or occupation before practical completion	56
	Practical completion	58
	Procedure at practical completion	60
	Failure to complete by the completion date	61
	Liquidated and ascertained damages	61
6	**Control of the works**	**67**
	Employer's agent	67
	Site manager and contractor's persons	67
	Principal contractor	68
	Principal designer	68
	Employer's obligations	68
	Information to be provided by the contractor	73
	Design submission procedure	73
	Schedule 1 procedure	74
	Employer's instructions	75
	Changes	78
	Goods, materials and workmanship	79
	Defective work	80
	Sub-contracted work	82
	Named sub-contractors	83
	Work not forming part of the contract/persons engaged by the employer	85
	Making good defects	85
7	**Sums properly due**	**89**
	Valuation of changes in the employer's requirements and provisional sum work	90
	Supplemental Provision 2: contractor's estimates	91
	Cost saving and value improvement	92
	Valuation under the valuation rules	92
	Measurable work – contract sum analysis	92
	Daywork – fair valuation	93
	Reimbursement of direct loss and/or expense	93
	Alternative procedure using Supplemental Provision 2 estimate	95

	Alternative procedure using Supplemental Provision 3 estimate	96
	Matters for which loss and expense can be claimed	96
	Fluctuations	99
8	**Payment**	**101**
	Timing of payments	102
	Ascertainment of amounts due	102
	Value of work	103
	Unfixed materials	104
	'Listed items'	106
	Costs and expenses due to suspension	106
	Other items in the gross valuation	106
	Deductions from the gross valuation	107
	Withheld percentage	107
	Advance payments and bond	107
	Payment procedure	107
	Deductions	108
	Employer's obligation to pay	109
	Contractor's position if the amount applied for is not paid	110
	Interim payment after practical completion	110
	Final payment	111
	Conclusive effect of final statement	111
9	**Indemnity and insurance**	**115**
	Injury to persons and damage to property caused by the negligence of the contractor	115
	Damage to property not caused by the negligence of the contractor	117
	Insurance of the works	118
	Action following damage to the works	119
	Terrorism cover	120
	Professional indemnity insurance	120
	Joint Fire Code	121
	Other insurance	121
10	**Default and termination**	**123**
	Repudiation or termination	123
	Termination by the employer	124
	Insolvency of the contractor	127
	Consequences of termination by the employer	128
	Termination by the contractor	130
	Consequences of termination by the contractor	131
	Termination by either the employer or the contractor	131
	Termination of named sub-contractor's employment	133
11	**Dispute resolution**	**135**
	Notification and negotiation	135
	Mediation	135
	Adjudication	136
	Arbitration	138
	Arbitration and adjudication	140
	Arbitration or litigation	140

References	**143**
Publications	143
Cases	143
Legislation	145
Clause Index	**147**
Subject Index	**153**

About this Guide

Design and build procurement is widely used in the UK, particularly for government-funded and larger projects, but recently there have been signs of it becoming increasingly popular on relatively small projects. The JCT standard form of contract for use with design and build procurement is widely used and is generally considered an industry benchmark.

When first published in 1981 it was a rather awkward adaption of the Standard Building Contract for use with traditional procurement, but later editions have refined the text and layout. The 2016 edition represents a big step forward in this process, whereby drafting has been simplified and rationalised in many areas, including the certification provisions, the stage payments option and the clauses covering action following damage to the works. Other key changes include incorporation of provisions relating to fair payment, the Public Contracts Regulations 2015 and the Freedom of Information Act 2000, modification of Insurance Option C to allow for more flexible solutions to insurance where work is done to existing buildings, the inclusion of performance bonds and parent company guarantees, and tightening up of the arrangements to secure provision of third party rights/warranties from sub-contractors. All of these will ensure that DB16 remains a popular choice for design and build projects.

This Guide is intended primarily for consultants, such as architects, quantity surveyors and project managers, who may be advising either the employer or the contractor. However, it would also be useful for the parties to the contract, and the author has recently come across examples of sub-contractors using it to help guide them where claims and disputes have arisen on design and build projects.

As with the other Guides in the series, this edition examines the form by topic, rather than on a clause-by-clause basis. After a general introduction to the form, which compares it with other JCT contracts, the Guide sets out some basic principles of design liability, with relevant case law and statutes. It then looks at the documents that form the contract package, including the employer's requirements and the contractor's proposals, followed by programming issues, control of the works, valuation and payment, insurance, termination and dispute resolution. Extensive use is made of diagrams and tables to further explain the form's procedures.

The author would like to thank her partner, Manos Stellakis, for his helpful comments and suggestions regarding the implications for consultants working in the context of design and build procurement.

1 Introduction to design and build procurement

1.1 In design and build procurement, the employer engages a single contractor to prepare and/or complete the design and to construct the works to meet the requirements of the employer. The employer may engage consultants, and the contractor will frequently sub-contract both the design and a large amount of the construction. At tender stage, the contractor will normally be given detailed information on the employer's requirements for the project, and sometimes an outline design. The contractor will normally submit a design proposal before the contract is signed, and will continue with the development of the design in parallel with construction.

1.2 Design and build can be contrasted with traditional procurement in that, with the traditional route, the contractor will normally take on little or no design responsibility. The employer will therefore appoint an architect and other consultants to develop the design and take the project through all the RIBA Plan of Work stages, including obtaining statutory permissions and preparing detailed production information, before tenders are invited from contractors.

1.3 The design and build route involves the contractor from an earlier stage. How early can vary considerably in practice, from situations where the contractor is involved soon after a decision to proceed with a building project has been taken, to situations where the employer engages a design team to prepare a detailed set of proposals, and the contractor is engaged to complete only the technical details and to construct the project.

1.4 The key advantages of the design and build route for the employer are that it provides a single point of responsibility for both design and construction, that there is a single price which covers design and construction, and that no further design information is required from the employer once the tender is let. In addition, as design proceeds in parallel with construction, the overall programme is normally shorter than when following the traditional procurement route. Where the contractor is involved in the design from an early stage, this will normally ensure that the developing design is efficient in terms of 'buildability', and some of the resultant savings will be passed on to the employer.

1.5 A key disadvantage for the employer is that there is little control over the details of the developing design, and therefore over the ultimate result. In contrast, under traditional procurement, the design is finalised to a high level of detail prior to tender, and there is full control over any subsequent developments. In addition, under most design and build contracts, any changes to the employer's requirements will result in heavy costs to the employer. Any assumption that this route gives better price certainty than traditional procurement should therefore be questioned. Under both routes, certainty will depend on a wide variety of factors, in particular how detailed the information is at tender stage and how restrained the employer is in requiring amendments.

1.6   In design and build procurement, advice given by the contractor to the employer will not be 'independent', as it would be in the case of professional advice under traditional procurement, where a consultant team is directly engaged by the employer, ensuring that the employer's interests are served. If the employer decides to engage independent consultants in the initial stages, and then to retain these during the construction phase, this will result in some duplication of fees as the contractor will have already included amounts for developing the design in its tender.

1.7   Design and build contracts first became popular in the UK during the 1970s, and the Joint Contracts Tribunal (JCT) produced the first standard form for design and build, 'With Contractor's Design' in 1981 (WCD81). Other JCT forms that may be used in a design and build context are the Major Project Construction Contract and the Constructing Excellence Contract. Other forms that are currently available and can be used in a design and build context are NEC3 and the FIDIC Yellow Book.

1.8   Design and build has remained popular since it was first introduced, accounting for around 40 per cent of construction projects in the UK (assessed by value, not number[1]). Its popularity somewhat levelled off during the 1990s, as in some cases it was found that the final result was not of the expected quality. In addition, newer forms of procurement, such as management contracting, attracted a lot of interest. However, its use has increased over the past few years, especially on larger projects. Design and build is used frequently in the context of public procurement, for example in relation to private finance initiative (PFI) arrangements, and has therefore undergone something of a revival, which shows no signs of abating.

1.9   In larger construction programmes a design and build contract may be linked to an umbrella framework agreement, or used in connection with single-project partnering. Originally developed in the private sector, this is increasingly used also in the public sector, and DB16 includes provisions that reflect this context. There is reference to a framework agreement, which would be used alongside the form. As an alternative, in cases where there is no partnering agreement, several of the optional supplemental provisions, such as collaborative working and key performance indicators, reflect a partnering ethos and could be used as a basis for single-project partnering.

1.10   Design and build often, although not necessarily, involves a two-stage tender process. The tender is split into two stages when insufficient information is available before the first tender period. The first stage will involve the submission of a tender sum and limited information regarding the design. The tender sum is based on preliminaries, and the contractor is asked to calculate a percentage overhead profit of the contract value. The second stage will normally involve negotiations regarding the tender and the submission of more detailed design proposals in order to reach an agreed contract sum.

1.11   As the contractor will normally be advising the employer during this phase, it is good practice to enter into a contract to cover the period. The JCT publishes a Pre-Construction Services Agreement (General Contractor) (PCSA), which is 'designed for appointing a contractor to carry out pre-construction services under a two-stage tender process', and a similar form for use with specialist contractors who may be providing advice during this stage (PCSA/SP). Although these are a useful way of engaging a company, it should be

---

[1] RICS and Davis Langdon, *Contracts in Use: A Survey of Building Contracts in Use during 2010* (London: RICS, (2012); also see the *NBS National Construction Contracts and Law Survey 2015* (Newcastle-upon-Tyne: NBS, 2015).

noted that they contain a provision that the contractor/specialist will not be liable for design advice given unless they are subsequently appointed for the construction phase, a proviso that could leave the employer in a difficult position should the advice later prove to have been negligent.

1.12 During the second stage of the tender process, the contractor is not in competition with any other contractors. This situation can make the process of establishing a contract sum problematic, but it is unwise to begin construction without a firm price in place (*Plymouth and South West Co-operative Society Ltd v Architecture Structure and Management Ltd; Trustees of Ampleforth Abbey Trust v Turner & Townsend Project Management Limited*). In addition to the general risk associated with proceeding on a letter of intent, it should be noted that certain obligations in DB16 depend on the contract being executed, for example to provide collateral warranties.

> ***Plymouth and South West Co-operative Society Ltd v Architecture Structure and Management Ltd* [2006] CILL 2366 (TCC)**
>
> Architecture Structure and Management Ltd (ASM) was the architect for this redevelopment project. The project was completed, but Plymouth and South West Co-operative Society Ltd (Plymco) claimed that an overspend of at least £2 million in excess of ASM's estimate of £6.3 million had occurred. Plymco claimed that much of that additional cost would have been avoided had ASM provided its services with reasonable skill and care, particularly with regard to the way in which it obtained tenders, arranged for the terms of the building contract, monitored the ongoing costs and operated cost-control procedures while work progressed. The project was let by a two-stage tender process, and using JCT98 With Approximate Quantities, without explaining to the client the risks involved. Following the first stage, negotiations did not proceed well, and by the end of the second stage 90 per cent of the value of the project was still covered by provisional sums. Although the client expressed concerns, the consultant recommended that the project should proceed. Altogether, approximately 7,500 architect's instructions were issued, leading to the excessive increase in costs. The court decided that the additional costs incurred, plus the potential cost savings which had been ignored, were recoverable from the architect.

> ***Trustees of Ampleforth Abbey Trust* v *Turner & Townsend Project Management Limited* [2012] EWHC 2137**
>
> The Trustees appointed Turner & Townsend to act as project manager for the building of a new school boarding house, which needed to be available for the start of term. To effect an early start on site, the project manager prepared a letter of intent while design and other issues were being finalised. A building contract was prepared but never entered into, and the work was carried out under eight letters of intent. The works were not completed on time and the Trustees wished to claim liquidated damages for the overrun. The draft building contract provided for liquidated damages at £50,000 per week, but the letters of intent stated that the contract terms would not be effective until it was executed. The court agreed that, in not arranging for the building contract to be executed, the project manager had acted negligently by failing to exercise reasonable skill and care and that the Trustees had suffered loss as a result.

1.13 Where single-stage tendering is used, all information regarding the design proposals and the pricing breakdown will be submitted with the tender, and a lump sum figure will be tendered, although it will often be subject to post-tender negotiation. Details of recommended procedures for design and build tendering can be found in the JCT's *Tendering Practice Note 2012* (JCT, 2012).

The architect's role

1.14 The architect may be involved in design and build procurement in one of several ways, e.g.:

- as a consultant to the employer during the initial stages of a project (advising on or formulating the employer's requirements, preparing an outline design and specification if needed, appraising potential contractors and assessing tenders);
- as an adviser to the employer during construction (advising on whether the developing design prepared by the contractor appears to meet the employer's requirements, on the effect of changes to the design and on the execution of the work);
- as the 'employer's agent' acting on behalf of the employer in respect of the construction contract;
- as a consultant to the contractor (preparing feasibility studies and design proposals, preparing contractor's proposals for tender and developing the design following tender to meet the employer's requirements).

1.15 Sometimes the architect transfers from the role of employer's consultant to that of contractor's consultant around the time that the design and build contact is entered into. This transfer is fraught with legal and practical difficulties. Two distinct ways of making this transfer are detailed below.

Novation

1.16 In this arrangement, the contract between employer and architect will be replaced by a contract on identical terms between the architect and the contractor. A simpler (although less accurate) way of describing the process is that the contractor will replace the employer as client under the original appointment. The contractor accepts all the obligations and liabilities that had formerly been the employer's under the appointment, and the architect's prior and future obligations/liabilities are now owed to the contractor. The architect will have no liability to the employer, including for any mistakes made earlier while working for the employer (unless a warranty is entered into). A deed of variation to the appointment is required to reflect this change, which should include any necessary or preferred alterations. All three parties enter into a novation agreement.

Consultant switch

1.17 In this arrangement, the original appointment with the employer as client is brought to an end and a new appointment is entered into between the architect and the contractor. A supplementary agreement between all three parties is necessary to permit this change. The architect will normally remain liable to the employer for any breach of duty under the earlier appointment, but will not be liable to the employer for any default in services performed for the contractor (unless an architect–employer warranty is entered into).

1.18 It is common practice for the term 'novation' to be used for both of the above methods of transfer, but this is an inaccurate use of the term, which has a specific legal meaning. Such inaccuracies can cause confusion in practice, since, as can be seen from the above descriptions, 'consultant switch' and 'novation' are quite different processes with different consequences for the architect in terms of liability. The difference was highlighted in the case of *Blyth and Blyth* v *Carillion*, where a contractor unsuccessfully sought to claim against the design team for mistakes made prior to a novation. Employers, consultants and, particularly, contractors should study the implications of this case carefully before entering into any consultant switch arrangement.

> *Blyth and Blyth* v *Carillion* (2001) 79 Con LR 142
>
> Consulting engineers Blyth and Blyth Ltd (Blyth) entered into a tripartite agreement, referred to as the 'novation agreement', with Carillion Construction Ltd (Carillion) and THI Leisure (Fountain Park) Ltd (THI) in relation to the design and construction of a leisure development building in Edinburgh. There was a deed of appointment between THI and Blyth, Section 6 of which empowered THI to instruct Blyth to enter into the novation agreement. The design and build contract between THI and Carillion was on an amended WCD81 form. Blyth brought an action against Carillion to claim payment of fees, and Carillion counterclaimed in respect of alleged breaches of contract by Blyth. This raised issues about the meaning and effect of the novation agreement, in particular in relation to alleged breaches occurring before the novation; one example being that, as a result of the engineer's inaccurate information regarding reinforcement bars, which was included in the employer's requirements, the contractor suffered losses when it eventually had to supply far more bars than it had anticipated. Under the amended terms of the contract, the contractor accepted the risk of inaccuracies in the requirements, and therefore could not claim these losses from the employer. The contractor therefore sought to claim them from the engineer. The court decided that the engineer was not liable to the contractor for losses due to breaches of its duty to the employer before the novation took place.

1.19 From the architect's point of view, making this transfer can involve many awkward legal and practical problems. It is often the employer's hope that, by arranging for this transfer to take place, it will retain control over the developing design through an ongoing link with the architect. However, the architect will in no sense be acting for the employer, as would be the case under traditional procurement. The architect should understand, and explain to the employer, that, from the moment the switch or novation takes place, the contractor takes over as the architect's client, and that, in future, any communication will be either through the contractor or approved by the contractor. For example, if the contractor sets out requirements for the developing design, this is a matter between the architect and the contractor, and should not be passed on to the employer, even if the architect is less than enthusiastic about these requirements. If the architect is aware that the employer has made a request or instructed a change, the architect has no right to incorporate this into the design until it is passed on by the contractor. In short, the architect will be putting the contractor's, and not the employer's, interests first. The JCT Practice Note CD/1A (see paragraph 2.5) put it in this way:

> In these circumstances the professional adviser(s) are responsible to their client (the Contractor) and are thus precluded from entering into any professional arrangement with the Employer for the purposes of the Contract. Any advice of a general nature that they may give will be given through the Contractor, or, if it is given direct, will be given on behalf of the Contractor. For the Contractor's professional advisers to act in any other way could cut across management and design responsibilities and thus prejudice the legal rights and obligations of the Employer and the Contractor.

1.20 Although it may not appear to be the case at first sight, the employer may be in a better position if it retains the architect to act as consultant and/or agent throughout the project, than if it attempts to retain control through the difficult and complex system described above. The architect, too, should be aware that he or she may be in a better position to influence the developing design by advising the employer on submissions from the contractor than by working for the contractor directly. Alternative arrangements, whereby the employer and contractor simultaneously engage the architect's firm to provide services in relation to the project are highly questionable, not least because the interests of the employer and the contractor are frequently opposed. Even if the device of a 'Chinese wall' is used, it is unlikely that the firm can avoid prioritising one party or the other in the advice it gives.

1.21   Whatever form the architect's involvement takes, it is very important that the terms of engagement are drawn up with great care. The architect should:

- ensure that the terms are acceptable to his or her professional indemnity (PI) insurers;
- avoid taking on any 'fitness for purpose' liability (see below);
- ensure that liability to the employer under any warranty is not greater than liability to the contractor;
- ensure that liability to the employer under any warranty is not greater than that owed by the contractor to the employer.

1.22   The RIBA Standard Agreement 2010 (2012 revision): Architect may be used as the basis for an agreement with the client to provide design or other services in design and build procurement (refer to the *Guide to RIBA Agreements 2010 (2012 revision)* for information).[2] The RIBA also publishes a supplementary Contractor's Design Services Schedule (SupCD-07), which covers the subject of the architect as contractor's consultant.

1.23   The Construction Industry Council (CIC) publishes a standard form of novation agreement, a standard form warranty and helpful guidance. When entering into such agreements, the terms should normally be checked with the consultant's PI insurers, and, in most instances, particularly when agreements are on non-standard forms, it is sensible to take additional legal advice. It is particularly important to check for obligations in relation to the timetable for provision of information to the contractor (see *Royal Brompton Hospital* v *Frederick Alexander Hammond* and *CFW Architects* v *Cowlin Construction Ltd*) and to identify any non-standard terms and conditions.

> *Royal Brompton Hospital National Health Service Trust* v *Frederick Alexander Hammond and others (No. 4)* [2000] BLR 75
>
> The Royal Brompton Hospital (RBH) engaged Frederick Alexander Hammond to undertake a £19 million construction project on a JCT80 standard form of contract. The contractor successfully claimed against RBH, including for losses suffered due to delays. RBH commenced proceedings against 16 defendants, who were all members of the professional team. A trial date was fixed to deal with a number of different issues, all of which were settled except for one relating to the consulting M&E engineers, Austen Associates Ltd (AA). The issue was whether AA was obliged to provide co-ordination and builder's work information to ensure that RBH complied with clause 5.4 of the main contract. The court decided that AA was under a duty to use reasonable skill and care in ensuring that the drawings were provided in time to enable the contractor to prepare its installation drawings, and thus to carry out and complete the works in accordance with the contract conditions.

---

[2] At the time of writing the RIBA is preparing a new suite of Professional Services Contracts which it intends to publish in 2017.

> **CFW Architects v Cowlin Construction Ltd (2006) 105 Con LR 116 TCC**
>
> Contractor Cowlin engaged architects CFW to carry out design work under a design and build contract for the Defence Housing Executive to build houses for service personnel. A contract was agreed, although not signed, and held to incorporate SFA/99 and a payment schedule. CFW delivered the drawings late, and when Cowlin did not pay, argued that Cowlin had repudiated the agreement. The court held that a term should be implied obliging CFW to supply drawings in accordance with the payment schedule in the main contract. Cowlin had not repudiated the agreement, but CFW had, and Cowlin was entitled to damages.

1.24   If the architect is to act as the employer's agent, the terms of appointment must give the architect at least the degree of authority that the employer is given under the design and build contract. It is important that the agency agreement is put into writing, again with legal advice.

1.25   Architects who are more accustomed to working in a traditional procurement environment often fall into one of several traps, which they must take care to avoid. One has already been mentioned – the tendency to continue to treat the employer as client after switching to work for the contractor. Another common problem arises where the architect, while retained by the employer, assumes an authority under the contract to which the architect is not entitled. If the architect does this, in some circumstances the contractor may be entitled to treat the architect as the employer's agent, particularly if the employer appears to endorse this behaviour. The employer will then be liable for the architect's actions, and may, in turn, seek to claim losses from the architect. A third problem occurs when, after being appointed as agent, the architect adopts an approach that would be appropriate for an independent administrator in a traditional contract. Not only are the roles quite different, but the specific tasks to be performed differ widely in many cases from the normal duties of a contract administrator. Altogether, the architect should proceed with great caution when appointed in design and build procurement, and make no assumptions about the actions he or she should be taking.

Some general principles of design liability

1.26   Before becoming involved in design and build procurement, it is obviously important to understand some of the basic principles of design liability. Standard forms of contract and appointment will often set out specific provisions regarding design liability, but these have to be understood in the legal context within which they operate. A key point is to determine whether any design liability incurred is a 'fitness for purpose' or 'reasonable skill and care' level of liability.

1.27   Under the Sale of Goods Act 1979 as amended by the Sale and Supply of Goods Act 1994, all contracts for the sale of goods contain implied terms that the goods sold will be of satisfactory quality. The Unfair Contract Terms Act 1977 stipulates that this requirement cannot be excluded from any consumer contract, and can only be excluded in other contracts insofar as it would be reasonable to do so. If parties have included terms which purport to exclude this liability, the terms will be void. Similarly, if the use to which the goods are to be put is made clear to the seller, the seller must supply goods suitable for that use, unless it is clear that the buyer is not relying on the seller's skill and judgment. So if, for example, a DIY enthusiast asks a builders' merchant for paint suitable for use on a bathroom ceiling, the merchant must supply suitable paint, regardless of what is written in the contract of sale. If, however, the buyer were to specify the exact type of paint, the seller would no longer be liable, as the buyer is not relying on the seller's advice.

1.28   Contracts for construction work are usually for 'work and materials' (as opposed to supply-only or install-only) and, as such, fall under the Supply of Goods and Services Act 1982. This implies similar terms to those described above in relation to any goods supplied under such a contract. Therefore, a contractor would normally be liable for providing materials fit for their intended purposes. If, however, an employer or architect specifies particular materials, the contractor would be relieved of this liability.

1.29   The obligation to supply goods or materials fit for their intended purpose would extend to a product or structure which a contractor had agreed to design and construct (*Viking Grain Storage Ltd* v *T H White*).[3] In all cases, the liability of the contractor will be strict; in other words, the contractor will be liable if the goods, element or structure is not fit for its intended use, irrespective of whether the contractor has exercised a reasonable level of skill and care in carrying out the design. This is a more onerous level of liability than that assumed by someone undertaking design services only, where they would normally be required to demonstrate that they had exercised the skill and care which could reasonably be expected of a competent member of their profession. In other words, if an employer can prove that a building designed and constructed by a contractor is defective, then this will normally be sufficient proof that there has been a breach of contract, whereas, in the case of a design professional, the employer would also have to prove that the professional had been negligent.

> ***Viking Grain Storage Ltd* v *T H White Installations Ltd* (1985) 33 BLR 103**
>
> Viking Grain entered into a contract with White to design and erect a grain drying and storage installation to handle 10,000 tonnes of grain. After it was complete, Viking commenced proceedings against the contractor claiming that, because of defects, the grain store was unfit for its intended use. The contractor, in its defence, claimed that there was no implied warranty in the contract that the finished product would be fit for purpose, and that the contractor's obligation was limited to the use of reasonable skill and care in carrying out the design. The judge decided that Viking had been relying on the contractor and, because of this reliance, there was an implied warranty that, not only the materials supplied, but the whole installation, should be fit for the required purpose. There could be no differentiation between reliance placed on the quality of the materials and on the design.

1.30   An architect involved with design and build should be careful not to take on a 'fitness for purpose' obligation, as it is unlikely that any PI insurance policy will cover this level of liability. In the case of *Greaves and Co. Contractors* v *Baynham Meikle*, an architect was found to have assumed such a liability, but it is important to note that this was implied on the particular facts and would not normally be implied as a matter of law.

---

[3] For a full discussion see Sarah Lupton, *Cornes and Lupton's Design Liability in Construction*, 5th edn (Chichester: Wiley-Blackwell, 2013).

> *Greaves and Co. Contractors* v *Baynham Meikle & Partners* (1975) 4 BLR 56
>
> Greaves entered into a design and build contract to provide a warehouse, factory and offices. The warehouse was to be used for the storage of oil. Greaves engaged Baynham Meikle, structural engineers, to design the structure of the warehouse, and told the engineers that the floor of the warehouse was to support the weight of fork-lift trucks carrying barrels of oil. Once the warehouse was brought into use, the floors began to crack and Greaves brought a claim against the engineers. It was decided at the first instance trial that the cracks were due to the vibrations caused by the trucks, and that the floor had not been designed to withstand these vibrations. The judge found for Greaves, and the engineers appealed. The appeal court confirmed that the engineers had been in breach of the clear but unexpressed intention that the engineers should design a warehouse suitable for the trucks, and therefore implied a 'fitness for purpose' obligation into the terms of engagement, rather than the lesser obligation to use due skill and care. The court emphasised, however, that the term had been implied due to the special facts of the case.

1.31   It should be noted that in clause 2.17.1 of DB16 the contractor's liability is limited to that of an architect or other professional designer, i.e. to the use of reasonable skill and care. This means that, in order to prove a breach, the employer would need to prove that the contractor had been negligent. If, for example, the contractor is required to design a heating system to heat rooms to a certain temperature, and, when installed, the system fails to do so, this fact alone would not be enough to prove that there had been a breach of contract. The employer would need to prove that the contractor had failed to use the skill and care expected of a professional person. However, this is dependent on the express terms being unamended.

1.32   Where the contractor carries out work in connection with a dwelling, including design work, this will be subject to the Defective Premises Act 1972. This states that 'a person taking on work for or in connection with the provision of a dwelling ... owes a duty ... to see that the work which he takes on is done in a workmanlike or, as the case may be, professional manner, with proper materials and so that as regards that work the dwelling will be fit for habitation when completed' (section 1(1)). This appears to be a strict liability and is owed to anyone acquiring an interest in the dwelling. Reference to this legislation and the resulting liability is made in DB16 (cl 2.17.2).

2 About DB16

2.1   The JCT Design and Build Contract (DB16) is intended for use on building projects where 'the Employer has defined his requirements and where the Contractor is not only to carry out the works, but also to complete the design for them in accordance with those requirements' (DB/G). It is applicable where the employer has issued a document to the contractor at tender stage stating the employer's requirements, and, in response, the contractor has submitted proposals for the design to the extent and detail required under the employer's requirements.

2.2   The form is published in only one version, for use by both private clients and local authorities. The contract particulars indicate that certain provisions (e.g. cl 4.6, provisions for advance payment and an associated bond) do not apply where the employer is a local or public authority. Special provisions are included which are required by or essential to its use by local authorities and public bodies, as discussed below at paragraph 2.18.

2.3   The form follows the normal JCT format of agreement, recitals, articles, contract particulars, attestation and conditions. The sixth recital makes reference to a framework agreement which may supplement the provisions of the contract. The form also includes 12 'Supplemental Provisions' under Schedule 2. These optional provisions are referred to in the seventh recital, and are incorporated by indicating whether or not they are to apply in the contract particulars (note that Part 1 of the Supplemental Provisions only applies if indicated, whereas Part 2 applies unless stated otherwise; supplemental provisions 11 and 12 apply only where the employer is a local or public authority, see footnote [9]). The operation of some of these provisions depends on the correct drawing up of the employer's requirements.

2.4   The form contains an advance payment bond (Schedule 6: Part 1), a bond in respect of payment for off-site goods and materials (Schedule 6: Part 2) and a bond in lieu of retention (Schedule 6: Part 3). The JCT collateral warranties from the contractor to a funder (CWa/F) and purchaser or tenant (CWa/P&T), which are published separately, may be used with DB16, as may the JCT collateral warranties from a sub-contractor to a funder (SCWa/F), purchaser or tenant (SCWa/P&T) and employer (SCWa/E).

2.5   The JCT publishes a guide for use with DB16 (the *Design and Build Contract Guide*, DB/G). This gives general guidance on the scope of the clauses and the changes since DB11. JCT Practice Note CD/1A gave guidance on the employer's requirements, the contractor's proposals, the contract sum analysis, value added tax and insurance to cover the contractor's design liability. CD/1B presented a commentary on the form, with notes on the contract sum analysis and the application of the formula adjustment for fluctuations. Despite the fact that the most recent editions of both practice notes were published as early as 1995, they still contain some very useful advice.

Key features

2.6　DB16 is a design and build 'lump sum' contract, under which the contractor is required to complete the design for, and to carry out the construction of, the work described briefly in the first recital and detailed in the contract documents, for the sum entered in Article 2, and to complete the work by the date entered in the contract particulars. It is important to note that the contractor is responsible for only that portion of the design which it actually completes, and not for the design as a whole.

2.7　The key differences between DB16 and traditional forms published by the JCT are, first, that the employer provides no further information to the contractor after the contract is entered into (the contractor has sole responsibility for completing the design) and, second, that no individual is appointed to exercise the functions of architect or contract administrator.

2.8　The contractor is responsible for constructing the works as described in the contract and for completing the design as necessary. The contract anticipates that the design could be at various stages of completion at tender stage. The second recital refers to documents termed the 'Employer's Requirements' and 'Contractor's Proposals', both of which form part of the contract documents. The requirements are sent out with the tender documents (and may contain an outline design) and the proposals are submitted with the tender.

2.9　The quality and quantity of work to be carried out is detailed in the employer's requirements and the contractor's proposals together with any further documents that the contractor prepares after appointment. The contractor is responsible for completing the design to meet the requirements, and for ensuring that it complies with statutory requirements. There is a requirement to submit the developing design proposals to the employer (cl 2.8 and Schedule 1).

2.10　The contractor's level of design liability is limited to that of a professional person (cl 2.17.1), therefore there is no strict duty to provide a building that meets the requirements set out in the employer's requirements, only to use due skill and care in preparing the design. If a higher level of liability is required, the employer may wish to consider amending this clause, although increasing the level of liability may result in higher tender prices.

2.11　The contract states what is to happen if there are discrepancies in either the requirements or the proposals. However, it does not expressly deal with a conflict between the two, i.e. if the proposals do not match the requirements. Although the employer is stated to be satisfied that the contractor's proposals appear to meet its requirements (third recital), it is likely that the requirements would take precedence (see paragraphs 3.39 and 3.40).

2.12　As no individual exercises the functions of the architect or contract administrator under DB16, the functions that might be ascribed to such a person under a traditional form are either omitted or modified, so that they can be undertaken either by the employer (which is the most usual situation) or the contractor.

2.13　The form provides for the employer to employ an agent to represent its interests, who could, of course, be an architect. The person named in the contract as agent will be 'stepping into the shoes' of the employer and will have authority to take the decisions and actions assigned to the employer under the contract. It should be noted that this is a very different role from that of contract administrator, as the agent is under no express obligation to act impartially, although it may be assumed that a duty of good faith will be implied.

2.14 The contractor tenders a lump sum and an analysis of that sum (the contract sum analysis). Payment to the contractor is made either at the end of agreed stages or at monthly intervals, following application by the contractor. In general terms, the payment will reflect the amount of work that has been properly completed in accordance with the terms of the contract up to the point of payment, plus the amount of design work that has been carried out, based on the breakdown in the contract sum analysis. The contract sum is adjusted in accordance with the conditions; for example, the provisions relating to instructions, loss and/or expense and fluctuations and, where relevant, this adjustment is based on the contract sum analysis. There are optional provisions for the use of bills of quantities.

2.15 All the requirements of the Housing Grants, Construction and Regeneration Act (HGCRA) 1996 as amended by the Local Democracy, Economic Development and Construction Act (LDEDCA) 2009 regarding payment and notices are, of course, incorporated into DB16, and the provisions regarding listed items and advance payments, which were introduced into JCT80 alongside the HGCRA 1996 requirements, are also included in DB16. The more recent revisions (since 2011) are discussed below at paragraphs 2.22 and 2.23.

2.16 The contractor may sub-contract the work, including the design for the work, to domestic sub-contractors with the written approval of the employer. The JCT publishes standard forms of contract for use with domestic sub-contractors, and the contractor is required to sub-let on these terms where appropriate. There are no provisions for nominating a sub-contractor but, under the Schedule 2 supplemental provisions, there are optional clauses allowing for a sub-contractor to be named. If these clauses are to be used, details must be included in the employer's requirements and the contractor is required to use the JCT named sub-contract forms for tendering and for the sub-contract. The contractor remains fully responsible for the performance of domestic and named sub-contractors, including any design they undertake, except that, in the event of termination of the employment of a named sub-contractor, the employer shares some of the risk for the additional costs of completing the work (see Table 2.1).

Table 2.1 Key responsibilities of the parties

The contractor must:

- complete the design (cl 2.1.1)
- carry out the work in conformity with the contract standards (cl 2.1.1) and with statutory requirements (cl 2.1.2) (note that this would include clearance of all outstanding planning and building control matters)
- complete on time (cl 2.3)
- appoint a full-time site manager (cl 3.2).

The contractor may:

- organise the work as it wishes
- object to unreasonable changes in the design sought by the employer (cl 3.5)
- seek payment for disturbance (cl 4.20) and changes (cl 5.2)
- with the consent of the employer, sub-contract the design or execution of the works (cl 3.3.1).

2.17   The form includes 12 optional supplemental provisions (Schedule 2), which allow it to be tailored to suit the parties' requirements. These include the use of named sub-contractors, contractors' estimates of the value of changes, acceleration and several provisions that reflect a partnering ethos, including collaborative working, cost savings and value improvements, and key performance indicators. It also allows for the use of a framework agreement, which might further enhance the partnering provisions, although such an agreement cannot override the provisions of the contract.

2.18   DB16 has many features required for public sector procurement, including 'Fair Payment' provisions, transparency and the use of BIM (building information modelling). These were originally published as a supplement to the JCT contracts, but have now been incorporated into the form with some further amendments. The Fair Payment provisions arise from the stated aims of the government in *Construction 2025*, which include equitable financial arrangements and certainty of payment throughout the supply chain. The aims are reflected in initiatives such as the Construction Supply Chain Payment Charter 2014, the HGCRA 1996 (as amended), the Late Payment of Commercial Debts Regulations 2013 and the Fair Payment Charter, as well as regulation 113 of the Public Contracts Regulations 2015. These require that the final date for payment should be 'no later than the end of a period of 30 days from the date on which the relevant invoice is regarded as valid and undisputed' (regulation 113(2)(a)), and that similar provisions are included in sub-contracts and sub-subcontracts. Under the charter, the value of work and materials supplied by all three tiers is to be assessed as at the same date. Adopting DB16 together with the appropriate JCT sub-contracts will ensure that the government requirements are met. The contract also includes provisions (Supplemental Provision 11) relevant to any employer that is subject to the Freedom of Information Act 2000 (which would include local authorities). This provides that the parties accept that the contract is not confidential, except for material that may be 'exempt' and which the employer has the discretion to determine. The Public Contracts Regulations 2015 also deal with corrupt practices and bribery, and breach of the statutory requirements is a ground for termination under clauses 8.6 and 8.11.3 of DB16. Furthermore, under Supplemental Provision 12 the contractor must include similar provisions in any sub-contract.

Deciding on DB16

2.19   DB16 is one of three JCT forms suitable if the contractor is undertaking a large part of the design. The two others, the Major Project Construction Contract and the Constructing Excellence Contract, are both intended for very large projects, and would normally require legal advice to prepare the tender and contract documents. For contracts where the contractor's design responsibility is restricted to discrete parts of the works, use of SBC16, with its provisions for a contractor's designed portion, should be considered. For smaller projects, where the contractor's design input is limited, ICD16 or MWD16 may be appropriate.

2.20   Care should be taken where the client is a consumer, particularly where the project involves providing a dwelling that the client intends to occupy. The HGCRA 1996 does not apply in this situation, and some of the ICD provisions may be considered unfair under the Unfair Terms in Consumer Contracts Regulations 1999 (SI 1999/2083) if not individually negotiated with the client prior to entering into the contract. In such situations, and where some design input is required from the contractor, the RIBA Domestic Building Contract would be particularly suitable.

Comparison with SBC16

2.21 For those who are familiar with SBC16, much of the text of DB16 will look surprisingly similar, given the considerable difference between the two methods of procurement that they cover. These similarities result from the JCT's sensible policy of adopting common wording wherever possible across forms. There are, of course, major differences between the forms. Some of the key similarities and differences between the forms are summarised in Table 2.2.

Table 2.2 Key comparisons between SBC16 and DB16

Key similarities:
- the provisions for a date for possession and date for completion, for extending time on the occurrence of 'Relevant Events' and for liquidated damages
- the provisions for carrying out the work in sections (i.e. phased work)
- the provisions for partial possession
- the provisions for practical completion and a rectification period
- the provisions for deduction of retention from payments
- the fluctuation provisions
- the insurance provisions
- the termination and dispute resolution provisions
- provision for purchaser's/tenant's and funder's rights from the contractor, either in the form of third party rights or by means of collateral warranties
- provision for collateral warranties from relevant sub-contractors, including in favour of the employer
- provision for withholding payment.

Key differences:
- the contractor is responsible for completing design and no further information is provided
- there are no provisions for 'listed sub-contractors'
- there are optional provisions for named sub-contractors
- there is no architect/contract administrator (decisions such as any extension of time to be awarded are ascribed to either the employer or the contractor)
- there is provision for a person to be named as employer's agent
- there is no clerk of works
- there is provision for the obtaining of planning permission and other approvals
- payment can be based on the value of work correctly carried out, or on the value of completed stages
- there is no certification of payment; the contractor makes application and the employer decides how much to pay.

Changes in the 2016 edition

2.22 Prior to the 2016 edition, the JCT had published one set of adjustments to the 2011 edition: Amendment 1 (March 2015), relating to the CDM Regulations 2015. The JCT had also published a Public Sector Supplement for use by local or public authorities. These changes are now incorporated in the 2016 edition.

2.23 The key 2016 changes are set out in Table 2.3 and can be summarised as follows:

- incorporation and updating of provisions from the JCT Public Sector Supplement relating to Fair Payment principles, transparency and BIM;

Table 2.3 Key changes

Clause	New/revised	Key changes
Contract particulars	revised	Supplemental provisions: site manager and bill of quantities are deleted, new Supplemental Provisions 11 (transparency) and 12 (Public Contract Regulations) added
Contract particulars	revised	Dates of interim valuation replace dates of payment application
Contract particulars	new	Entries in respect of performance bonds and guarantee
Contract particulars	revised	Third party rights and warranty information to be set out in separate 'Rights Particulars'
1.1	revised	'BIM Protocol', 'C.1 Replacement Schedule', 'Consultants', 'Design Submission Procedure', 'Employer Rights', 'Existing Structures', 'Final Payment Notice', 'Interim Payment Application', 'Interim Valuation Date', 'Local or Public Authority', 'Named Sub-Contractor', 'PC Regulations', 'Principal Designer', 'Rights Particulars' and 'Works Insurance Policy' added to defined terms
1.4	revised	References to documents to include information in a format set out in any BIM protocol
1.8.2	revised	Effect of conclusiveness in relation to dispute resolution clarified
1.10	new	All consents and approvals not to be unreasonably delayed or withheld
2.38.3	new	Provision for licence to be assignable
3.2	new	Provision for site manager added
3.4.2.5	new	Sub-contracts to provide for granting of third party rights or execution of warranties
3.4.3	new	Sub-contracts to provide for supply of information and grant of licences in relation to the BIM protocol
3.16	revised	CDM Regulations 2015
4.2	revised	General adjustments simplified
4.7	revised	Interim payment intervals the same for period payments and stage payments
4.9.1	revised	Final payment 14 days from due date (not 28 as in DB11)
4.19	revised	Loss and expense application procedure simplified
6.2, 6.3	revised	Contractor's liability for damage due to its negligence clarified
6.13, 6.14	revised	Procedures for making good after damage re-drafted and simplified; included in main contract text, not in schedules
7.3	new	New provisions for performance bonds and guarantees
7E	revised	Warranties/third party rights from sub-contractors redrafted and clarified
8.11.3	new	Right to terminate in relation to the Public Contracts Regulations 2015 added

Table 2.3 Key changes – Continued

Clause	New/revised	Key changes
Schedule 2	deletion	Supplemental provisions for a site manager and bills of quantities removed
Schedule 2	new	Supplemental provisions for transparency and the Public Contracts Regulations 2015 added
Schedule 3	new	Provision for a 'C.1 Replacement Schedule' added

- amendments relating to the CDM Regulations 2015;
- reference made to various provisions of the Public Contracts Regulations 2015;
- changes in respect of payment, designed to reflect Fair Payment principles and to simplify and consolidate the payment provisions;
- payment is monthly, whether using the 'periodic' or 'stage payment' option;
- removal of Fluctuations Options B and C – these are now published separately;
- the inclusion of performance bonds and parent company guarantees;
- tightening up of the arrangements to secure provision of third party rights/warranties from sub-contractors;
- simplification and rationalisation of drafting in many areas; for example, calculation of amounts due and provisions for rectification following damage covered by insurance.

3 Documents

3.1 DB16 defines the 'Contract Documents' as being the agreement and conditions (i.e. the printed form), the employer's requirements, the contractor's proposals, the contract sum analysis and '(where applicable) the BIM Protocol': these documents comprise the contract to which the parties are bound.

3.2 The articles of agreement and contract particulars must be completed carefully. All the information inserted must reflect that issued at tender stage, or agreed during post-tender negotiations. Ideally, the formal contract documents should be executed before the project commences on site. Normally, a contract is formed if there is a clear acceptance of a firm offer.[1] The contract, once executed, will supersede any conflicting provisions in the accepted tender and will apply retrospectively (*Tameside Metropolitan BC* v *Barlow Securities*).

> *Tameside Metropolitan Borough Council* v *Barlow Securities Group Services Limited* [2001] BLR 113
>
> Under JCT63 Local Authorities, Barlow Securities was contracted to build 106 houses for Tameside. A revised tender was submitted in September 1982 and work started in October 1982. By the time the contract was executed, 80 per cent of building work had been completed, and two certificates of practical completion were issued relating to seven of the houses in December 1983 and January 1994. Practical completion of the last houses was certified in October 1984. The retention was released under an interim certificate in October 1987. Barlow Securities did not submit any final account, although at a meeting in 1988 the final account was discussed. Defects appeared in 1995, and Tameside issued a writ on 9 February 1996. It was agreed between the parties that a binding agreement had been reached before work had started, and the only difference between the agreement and the executed contract was that the contract was under seal. It was found that there was no clear and unequivocal representation by Tameside that it would not rely on its rights in respect of defects. Time began to run in respect of the defects from the dates of practical completion; the first seven houses were therefore time barred. Tameside was not prevented from bringing the claim by failure to issue a final certificate.

3.3 Supplemental Provisions 1 to 10 are incorporated by deleting the words 'does not apply' against each required provision in the Schedule 2 entry in the contract particulars. Any supplemental provisions that are not required should be deleted, and an appropriate entry made in the conditions. Any amendments to the terms themselves, or additional conditions, should be incorporated by means of an additional article.

3.4 The form contains an attestation for formal execution of the contract. Execution 'under hand' is a straightforward matter, whereby the parties or their authorised representatives

---

[1] See, for example, Stephen Furst and Vivien Ramsey (eds.), *Keating on Construction Contracts*, 10th edn (London: Sweet & Maxwell, 2016) or Peter Aeberli, *Focus on Construction Contract Formation* (London: RIBA Publishing, 2003).

sign in the presence of witnesses. There are four alternative methods of execution as a deed. For individuals, a signature in the presence of a witness is required. For companies, either two signatories are required, one of which must be a director and the other either another director or the company secretary or, instead of signing, the company's seal may be affixed in the presence of these two people. A third option, since April 2008, is that a single director's signature is sufficient, provided that it is witnessed (Companies Act 2006, section 44(2)(b)). This form of execution is included in DB16. The form does not require any of the other contract documents to be signed, and as they are all to be uniquely identified in the contract particulars with exact titles, revisions numbers, etc., this should not be necessary but, to avoid any possible doubt in practice, a set is often signed.

Employer's requirements

3.5 The contract does not stipulate any format for the employer's requirements. It does, however, refer to and rely on the requirements in many places (e.g. under Schedule 1 the contractor is to provide information in the format set out in the employer's requirements), and it is therefore important that the requirements address all these issues.

3.6 In broad terms, the document sets out the employer's requirements for the completed building. It should be prepared carefully and on the assumption that there will be no changes to the requirements once the contract is let, for although the contract contains provisions whereby a change can be instructed, such changes may result in additional costs to the employer, and are subject to the consent of the contractor. Significant changes will negate many of the perceived advantages of this form, and if these are anticipated it may be more appropriate to consider using another form.

3.7 The requirements may be in a very summary format; for example, simply giving a schedule of accommodation. However, it is likely that, in practice, the information will be much more detailed and will include a detailed specification, in either prescriptive or performance terms, or, in all probability, will involve a mixture of the two. It may also include, for example, schedules of finishes, fittings and services requirements. It could also include schematic layouts or outline designs. In essence, it acts as a brief.[2] Where descriptive or performance specifications are included, these should be accurate. Use of the phrases 'to be to the employer's approval' or 'to be approved' should be avoided at all costs (see paragraph 8.45).[3]

3.8 The contract does not refer to the use of bills of quantities (this is a change since DB11). It sometimes occurs that only a part of the works will be designed in detail by the employer's design team, in which case it may be helpful to use a bill for that part only. If bills are included, amendments to the from will be needed to indicate how they will used; for example, will they contribute to the valuing of changes, and will they act as a basis for the assessment of periodic payments?

3.9 The document should state whether any development permissions have been obtained, including planning permission. If there are conditions attached to any approvals these

---

[2] Advice on briefing is outside the scope of this Guide, but reference could be made to Paul Fletcher and Hilary Satchwell, *Briefing: A Practical Guide to RIBA Plan of Work 2013 Stages 7, 0 and 1* (London: RIBA Publishing, 2015).
[3] Guidance on the preparation and use of performance specifications can be found in the *JCT Guide to the Use of Performance Specifications* (London: RIBA Publishing, 2001).

should be set out. If planning permission has still to be obtained, the requirements should specify who will take responsibility for this (although it should be noted that this will be the contractor's responsibility unless stated otherwise). It should also be stated, with reference to clause 2.15.2.2 of the contract, whether any amendments needed to comply with planning requirements are to be treated as a 'Change'. In this case, the amendments would be the employer's risk, unless the requirements stipulate otherwise.

3.10 The requirements should also include information on any covenants or easements relating to the site, and the extent to which the contractor is to base its proposals on site information provided by the employer, or is to make its own investigations. The requirements should stipulate any constraints which the employer wishes to impose on the use of the site or its facilities.

3.11 The requirements should include all the items to be entered in the contract particulars, as well as dealing with all the issues covered in Table 3.1 (which lists the clauses in the conditions that refer to matters that could be detailed in the employer's requirements), and

Table 3.1 Matters relating to the form that should be covered in the employer's requirements

Clause	
Fifth recital	Any division of the works into sections
2.1.2	The extent to which the requirements are to be taken as conforming to statutory requirements
2.2.3	Submission of samples, including timing and procedure for employer's comments
2.6	Information relating to work to be carried out by persons engaged directly by the employer
2.8	Any special procedure for submission of drawings and other information (over and above those in Schedule 1), including timing
2.9	The site boundaries
2.15.2.2	Whether amendments to the contractor's proposals which are necessary to conform with decisions of the statutory authorities after the base date are to be treated as an instruction requiring a change
2.18	Any provisional sums allowed for statutory fees
2.37	Detailed requirements in respect of as-built drawings
3.11	Provisional sums
4.7.4	Supporting detail required with applications for interim payment
6.5.1	Whether insurance against non-negligent damage to property is required
Schedule 1:	
1	Format for submission of drawings and other information
Schedule 2:	
1.1	Details of persons named as sub-contractors, and the work to be carried out by them
9.1	Performance indicators against which the contractor's performance is to be measured

which are not also covered by entries in the contract particulars. One of the most important inclusions is to stipulate exactly in what form the contractor's proposals should be submitted, and what they should include. This is essential so that the employer can make a clear assessment of the submitted tenders.

3.12　It is also very important that the requirements should specify the drawings and other design information (the 'Contractor's Design Documents') to be submitted by the contractor following acceptance of tender, and a programme for their submission. The purpose of this requirement is to control the scope, format and timing of the submission of design documents for review. For example, it should protect the employer from being overwhelmed by design documents at an inconvenient time, or from being presented with design documents to review for key elements in the absence of information on other related aspects of the design. It is likely that the programme will be the subject of negotiation at tender, as it is important that any programme in the requirements will also meet the contractor's needs in terms of developing the design at a rate which will support its intended construction programme. It would also be wise to set out the information required to be submitted at practical completion, such as 'As-built Drawings', otherwise the contractor's obligation is to provide such information 'as the Employer may reasonably require' (cl 2.37).

Contractor's proposals

3.13　The contractor's proposals should be in the format and contain the information stipulated in the employer's requirements. These may request that various documents are provided, including drawings, specifications, schedules, programmes, method statements, etc.

3.14　The contractor should raise matters relating to the contract data so that any necessary outstanding decisions can be made by the employer. The proposals should indicate clearly any areas of conflict in the requirements, and any instances where the contractor has found it necessary to amend or amplify the brief. The contract does not allow for the inclusion of provisional sums in the proposals, only in the requirements (cl 1.1 and 3.11), so if the contractor wishes to cover any part of the proposals with a provisional sum, then it should inform the employer so that the requirements can be amended.

Contract sum analysis

3.15　The contract does not prescribe a format for the contract sum analysis. It would therefore be advisable to stipulate an acceptable format in the employer's requirements. (It would not be unreasonable, for example, for the contractor to be asked to prepare a full bill of quantities, although in practice this would be unlikely on smaller projects.) The contract requires that the document is used for assessing the value of employer-instructed changes, for work which was covered by a provisional sum in the requirements and, where Fluctuations Option C is applicable, to enable the operation of this clause in accordance with the formula rules. The conditions do not require that the document is used to assess the value of work carried out, etc., to be included in periodic payments, but it would normally be used by the contractor to prepare applications for payment, and by the employer in checking such applications.

3.16 It is also important that a clear basis for assessing the value of design work is included. If the employer has exercised its right under Supplemental Provision 1 (Schedule 2) to name sub-contractors, then the contractor should also be required to itemise the parts of the contract sum which relate to these.

BIM and other protocols

3.17 If building information modelling (BIM) is to be used on the project, it will be important that the parties agree many matters to do with how the model will be prepared and managed, such as format, communication methods, timing, the detecting and resolving of problems, copyright, use following completion, etc. The contract allows for the use of a 'BIM Protocol', and the parties should adopt a standard form protocol or prepare a bespoke one for the project. At the time of writing, the only standard form available is the *Building Information Model (BIM) Protocol* published by the Construction Industry Council (CIC, 2013 available free from the CIC website). The protocol to be used should be identified in the contract particulars (cl 1.1). Clause 1.4.6 states that 'references to documents shall, where there is a BIM Protocol or other protocol relating to the supply of documents or other information, be deemed to include information in a form or medium conforming to that protocol'. (Note that cl 1.1 refers only to a BIM protocol; if some other protocol is needed then this would require some minor amendments.)

Health and safety documents

3.18 The employer and contractor are required to comply with the Construction (Design and Management) (CDM) Regulations 2015 (cl 3.16). A key element of the Regulations is the employer's duty to appoint a principal designer and a principal contractor (regulation 5); on most projects using DB16 the contractor will fulfil both roles, at least during the construction phase (see discussion in paragraph 4.28). In addition to the general obligation to comply with the Regulations, clauses 3.16.1 to 3.16.5 refer to various specific obligations. Particularly important in regard to documentation are matters relating to the construction phase plan and the health and safety file.

3.19 The construction phase plan is not a contract document under DB16, and the recitals make no mention of it having been prepared and given to the contractor at the time of tender. However, the employer (and the principal designer, if not the contractor) must provide the contractor with pre-construction information (regulations 4(4) and 12(3)), which should be sent out with the tender documents. Where the contractor is the principal contractor, it must ensure that the construction phase plan is prepared before setting up the construction site (regulation 12(1)); compliance with this is required under clause 3.16.3. To avoid uncertainty, it is advisable to require that this document be submitted by the contractor well in advance of the start of work on site. Following commencement, the contractor must ensure that the plan is reviewed and updated on a regular basis (regulation 12(4)).

3.20 Under the Regulations the health and safety file is principally a matter for the principal designer, who will compile it (regulation 12(5) and (6)), but there is a requirement on the contractor to provide information for the file (regulation 12(7)). Under DB16 the contractor, where it is the principal designer, must prepare the health and safety file and deliver it to the employer (cl 3.16.2). In addition, clause 2.27 requires the contractor to have complied with all its CDM duties with respect to the supply of documents and information before a statement of practical completion is issued.

Bonds

3.21 DB16 refers to several types of bond: an advance payment bond, a bond in respect of payment for off-site materials and/or goods, a retention bond and a performance bond. Where bonds are required, these must be arranged by the contractor and, as all bonds are optional, it must be made clear to the contractor at tender stage if any will be required. The first is normally required where an advance payment is to be made to the contractor under clause 4.6. The second is for use where it has been agreed that certain materials or goods will be paid for in advance of their being brought onto site (cl 4.15). The retention bond is used where a contractor is required to provide security in lieu of the normal deduction of retention from amounts included in interim payments. Only the second of the bonds may be used where the employer is a local or public authority. Terms for each of these bonds have been agreed between the British Bankers' Association and the JCT and are included in the form at Schedule 6. Performance bonds are discussed under paragraph 3.46 – if one is required, or if any other type of bond is required, this must have been made clear at tender stage, and the form of bond must be given to the contractor before the contract is entered into. In practice, the alternative terms should be sent out with the tender documents, so that the contractor can provide for them in the tender figure.

Sub-contract documents

Domestic sub-contracts

3.22 The JCT publishes a standard form of sub-contract for use with domestic sub-contracts under DB16, comprising an agreement (DBSub/A) and conditions (DBSub/C). The contractor is required to sub-let on these terms where appropriate (cl 3.4). There are also restrictions on the terms that may be agreed in any sub-contract. These are set out in clause 3.4 of DB16, which requires that particular conditions relating to termination, ownership of unfixed goods and materials, access to workshops, the CDM Regulations, the right to interest on unpaid amounts properly due to the sub-contractor, third party rights and warranties are included in all domestic sub-contracts. The sub-contract should also, of course, comply with the requirements of the HGCRA 1996 (as amended).

3.23 The provisions for a named sub-contractor under Schedule 2 paragraph 1 do not require that any particular form of sub-contract is used. However, the clause 3.4 requirements, as set out above, would apply to contracts with named sub-contractors. The contractor is also required to include the provision set out in Schedule 1 paragraph 1.5 (discussed later at paragraph 10.36). The employer may wish to stipulate in addition that specific terms are used for all named sub-contracts. It may be particularly important, where the sub-contractor is undertaking a significant element of design, to arrange for a collateral warranty, in order to allow the employer a means of claiming against the named sub-contractor should that part of the design fail (see paragraph 3.62). If the contractor should become insolvent and there is no warranty in place, the employer will not be able to recover its losses.

Use of documents

Interpretation, definitions

3.24 Clause 1.1 sets out definitions of terms that are used throughout the contract. Many of these are in common with other JCT forms, but some are unique to DB16, namely:

- change (refers to clause 5.1);
- contractor's proposals (refers to the second recital and contract particulars);
- development control requirements (any statutory provision and any decision of a relevant authority thereunder which control the right to develop the site);
- employer's agent (refers to Article 3);
- employer's requirements (refers to the first recital and contract particulars);
- final statement and employer's final statement (refer to clauses 1.8 and 4.24).

3.25 As can be seen from the examples above, further and more detailed definitions are frequently embodied in the text of clauses; for example, 'All Risks Insurance', 'Joint Names Policy' and 'Specified Perils' are defined at clause 6.8. Some derive from the HGCRA 1996 (as amended) and restate its requirements relating to the calculation of periods of days and the serving of notices (cl 1.5 and 1.7). Clause 1.7.1 requires that any notice or other communication expressly referred to in the agreement or conditions must be in writing. Clause 1.7.2 then states 'Subject to clause 1.7.4, each such notice or other communication and any documents to be supplied may or (where so required) shall be sent or transmitted by the means (electronic or otherwise) and in such format as the Parties have agreed or may from time to time agree in writing for the purposes of this Contract'. This would allow the parties to agree that all communication is by electronic means. It should be noted, however, that the contract sets out specific requirements for notices in some situations (for example, termination), which refer to clause 1.7.4, whereby all notices must be 'delivered by hand or sent by Recorded Signed for or Special Delivery post'.

3.26 Clause 1.6 was introduced through Amendment 2 to WCD98 in response to the coming into effect of the Contracts (Rights of Third Parties) Act 1999. In broad terms, this Act created rights for a person not party to a contract to bring an action for breach of a contract, where that contract expressly gave a benefit, or purported to give a benefit, to that person. This clause prevents any such claims being brought, by making it clear that the contract confers no rights on third parties, other than any specifically set out in clauses 7A and 7B of the contract (see 3.50).

3.27 Clause 1.11 states that the law of the contract will be English law. This would apply even if the contract was signed, or the work was carried out, in another jurisdiction.

Priority of contract documents

3.28 Clause 1.3 states that nothing contained in the employer's requirements or the contractor's proposals or the contract sum analysis, nor anything in any framework agreement, shall override or modify the agreement or the conditions. Were this clause not included, the position under common law would be the reverse. In other words, anything stated to be specifically agreed and included in a document would normally override any standard provisions in a printed form.

3.29   If the parties wish to agree to any special terms that differ in any way from the printed conditions, then amendments should be made to the form. This could be done either by amending the clauses themselves, or by inserting an additional article referring to the special terms, which should be appended to the form. The article could take a similar form to that used by the JCT to incorporate separately published amendments. However, attempting to amend standard forms without expert advice is very unwise as the consequential effects are difficult to predict. Deleting clause 1.3 may be particularly unwise as it may have unintended effects on other parts of the contract.

3.30   It should be noted that the CIC BIM Protocol contains a priority clause, which states that in the case of conflict with other contract document, the CIC protocol will prevail (paragraph 2.1). There is therefore a potential clash between DB16 and the protocol if they are used together. In reality, this should not present a significant problem with respect to DB16; provided that the protocol is completed in such a way as to deal solely with BIM-related matters, as the only clauses in DB16 that mention BIM are simply referring to the protocol, which will supplement rather than conflict with the form. However, there is potential for conflict between the protocol and matters set out in the employer's requirements, for example in relation to document provision. Although it is likely that the protocol would take precedence, the 'discrepancies and divergences' clauses (cl 2.10 to 2.14) do not specifically mention the BIM protocol, therefore it is not clear how the conflict will be resolved. It may be sensible for the parties to give this some thought, and to set up a clear document hierarchy between all relevant documents and a procedure for notification and resolution should conflicts be discovered.

Discrepancies and errors

3.31   It is preferable that all inconsistencies, errors or omissions within or between contract documents are corrected before the contract is entered into. Footnote [3] to the third recital, states that 'Where the Employer has accepted a divergence from his requirements in the proposals submitted by the Contractor, the divergence should be removed by amending the Employer's Requirements before the Contract is executed'. In many cases the later resolution of errors and inaccuracies will result in additional costs to the employer and, in some cases, in an extension of time.

3.32   The employer is obliged to issue instructions to deal with any divergence between the employer's requirements and the definition of the site boundary to be given under clause 2.9. The instruction 'shall be treated as a Change' (cl 2.10.1), and is therefore to be valued under clause 5.6. It is also listed as a relevant event (cl 2.26.1), and under clause 8.11.1.2 as a ground for termination by the contractor, should the instruction result in a suspension of the works. It is not specifically referred to under clause 4.19 as a matter giving rise to a claim for direct loss and/or expense, and it is suggested it should not be treated as falling under the remit of clause 4.21.1.

3.33   The employer must also take steps to deal with any inadequacy that is discovered in the design contained in the employer's requirements (cl 2.12), if this is not already covered in the proposals. Such 'correction, alteration or modification' shall be treated as a change (cl 2.12.2). This would be the case however the correction was effected, whether or not it was covered by an instruction, but in practice it is wise to cover all such corrections by an instruction to ensure clarity.

Between documents

3.34 The contractor is required to notify the employer if it finds any discrepancies or divergences 'in or between' the employer's requirements, the contractor's proposals and other design documents, and any instructions (other than one requiring a change) (cl 2.13). The employer must also issue instructions regarding the discrepancy, and the contract sets out specific procedures for particular categories of discrepancy or divergence.

Contractor's proposals

3.35 Where there is a discrepancy 'within or between the Contractor's Proposals and/or other Contractor's Design Documents' the contractor must inform the employer of its proposed amendment to deal with the discrepancy. The contractor is obliged to accept the employer's decision and comply at no cost to the employer (cl 2.14.1). If there were to be undue delay by the employer in reaching a decision, however, then this would constitute grounds for an extension of time (cl 2.26.1 or 2.26.6) and loss and/or expense (cl 4.21.2 or 4.21.5).

Employer's requirements

3.36 Where there is a discrepancy within the employer's requirements, or a discrepancy between the requirements and any change, the contract states that if the contractor's proposals deal with the discrepancy then they will prevail (cl 2.14.2). The discrepancy between the requirements and any change refers to inadvertent problems resulting from the effect of a change, rather than intended alterations to the particular part of the requirements at which the change was aimed. If the employer decides that it dislikes the solution in the contractor's proposals, and would prefer some other solution, this would have to be instructed as a change.

3.37 If the proposals do not deal with the discrepancy, the contractor is required to inform the employer of its proposal for dealing with the discrepancy, and the employer must either agree or decide on alternative measures and, in either case, notify the contractor in writing (cl 2.14.2). The acceptance or notification is to be 'treated as a Change', which would result in it being valued under clause 5.6. It would also constitute grounds for an extension of time under clause 2.26.2.1, for loss and/or expense under clause 4.21.1, and for termination under clause 8.11.1.2, in the unlikely event that it causes a suspension. As stated in paragraph 3.35, if the employer failed to reach a decision within a reasonable time, this would be grounds for an extension of time (cl 2.26.1) and a claim for loss and/or expense (cl 4.21.2).

3.38 It should be noted that, with respect to the clause 2.10 and 2.14 divergences, although the parties are required to notify each other immediately regarding any divergences they discover, there is no express obligation for the contractor to search out divergences. It is likely, however, that an obligation to identify at least obvious errors would be implied as part of the normal duty to use reasonable skill and care.

Employer's requirements and contractor's proposals

3.39   The contract does not deal with the situation where a divergence between the employer's requirements and the contractor's proposals is discovered after the contract is entered into. The third recital states that 'the Employer has examined the Contractor's Proposals and, subject to the Conditions, is satisfied that they appear to meet the Employer's Requirements'.

3.40   The recital is somewhat problematic from the point of view of the employer in that it appears to give precedence to the contractor's proposals and the contract sum analysis. It is suggested, however, that the contractor is obliged to meet the employer's requirements, even if an aspect of its accepted proposals does not initially comply. The overall structure of the conditions is that the contractor must provide a design that meets the employer's requirements. In addition, as the employer has no power to amend the contractor's proposals, there would be no means of effecting any change if this document always took precedence.[4] Nevertheless, the recital is often deleted in practice to avoid any scope for argument.

3.41   A footnote to the recital explains that if a discrepancy is discovered before the contract documents are executed, and the employer is prepared to accept the contractor's proposals, then the employer's requirements should be amended accordingly. Even if discovered afterwards, the matter can be remedied by a change instruction, provided the employer is happy with the contractor's proposal. Where the employer would prefer the version set out in its requirements, there is no need for any change, although it might be sensible to confirm the position with the contractor in writing.

Divergences from statutory requirements

3.42   The contractor and the employer are required to notify one another of any discrepancy or divergence between the employer's requirements or the contractor's proposals or other contractor's design documents, and any statutory requirement as defined under clause 1.1 (cl 2.15.1). The contractor must inform the employer of its proposed amendment and the contract requires the employer to 'note the amendment on the Contract Documents', i.e. the executed contract documents (cl 2.15.1). The contract states that the contractor must comply at no extra cost to the employer, unless the discrepancy results from a change in statutory requirements since the base date, in which case the instruction is treated as a change (cl 2.15.2.1). In addition, unless precluded by the employer's requirements, any modification to the proposals required by the terms of an approval is also to be treated as a change. This applies to permissions and approvals by statutory authorities relating to 'Development Control Requirements' as defined in the contract, and the approval must have been given after the base date. Finally, if the parts of the requirements which specifically state that they comply with statutory requirements are found not to do so, the employer must issue instructions requiring a change to remedy the situation. The effect of this clause is that the costs will be borne by the contractor in situations where the divergence is between the contractor's proposals and statute, and in limited circumstances where it is between the employer's requirements and statute.

---

[4]   For a detailed discussion, see David Chappell, *The JCT Design and Build Contract 2011* (Oxford: Blackwell, 2014).

Custody and control of documents

3.43 The employer's requirements, the contractor's proposals and the contract sum analysis remain in the custody of the employer, and must be available for inspection by the contractor at all reasonable times (cl 2.7.1). The contractor must be provided with one certified copy of the certified contract documents, unless the BIM protocol or other communication protocol states otherwise (cl 2.7.2). This should be done immediately after the execution of the contract. Although it is frequently done in practice, there is no need to sign two copies of the contract. It is easy to make minor mistakes when filling out two copies of the form, and it is safer to have one definitive set of contract documents, with certified copies made as required.

3.44 The contractor is obliged to supply the employer with two copies of each of the contractor's design documents, unless stated otherwise in the employer's requirements or contractor's proposals (cl 2.8, Schedule 1). The contractor is required to keep copies of the contract documents and the contractor's design documents on site at all times (cl 2.7.3).

3.45 Clause 2.7.4 stipulates that the copies of the contract documents and the contractor's design documents must not be divulged or used for any purpose other than in relation to the contract, and that the same restriction applies to 'any confidential information of the other party', but allows the employer a limited right to use any document supplied by the contractor for the purpose of maintenance, use, repair, advertisement, letting or sale of the works. Where the employer is a local or public authority, the obligations of confidentiality are subject to the specific requirements of Schedule 2: Supplemental Provision 11 (cl 2.7.5).

Performance bonds and guarantees

3.46 It is becoming increasingly common on larger projects for employers to ask for a performance bond from the contractor. This is a device to protect the employer from the risk of the contractor failing to perform the contract, and can be provided by a bank or an insurance company. An alterative is to ask the contractor to provide a guarantee from its parent company. If either of these is required, details should be included in the tender documents, and entered into the contract particulars (cl 7.3). DB16 does not contain a model form of bond, so full details of the document will be needed and the bond or guarantee must be provided on execution of the contract (cl 7.3).

Assignment and third party rights

Assignment

3.47 The right of a subsequent purchaser to bring an action against the builder of their property, with whom they have had no contractual relationship, could be of considerable value. The employer in a construction contract might therefore wish to assign this right to such other person as may acquire an interest in the property.

3.48 A contractual right can be regarded as a personal right of property, and in property law it is classified as a 'chose in action'. Choses in action can be assigned under the Law of Property Act 1925, provided that the requirements of section 136 of the Act are followed.

It is important to note that only contractual rights, termed 'the benefit of a contract', can be assigned and not obligations. So if, for example, A enters into a contract with B whereby A agrees to carry out some building work, and B agrees to pay A £100 for the work, A can assign the right to claim the £100 to C, but not the obligation to carry out the work. The right to pursue a debt or claim is assignable to C without B's consent, provided that B is notified as required by section 136. The obligation to carry out the work, however, could only be transferred to C with the agreement of all three parties (often termed 'novation').

3.49 DB16 contains express provisions which limit the scope for assigning contractual rights. Clause 7.1 states that neither the employer nor the contractor may 'assign this Contract or any rights thereunder' without the written consent of the other. Assignment without consent of the other party is grounds for termination (cl 8.4.1.4 and 8.9.1.2). There is one exception, however, to the prohibition on assignment: if clause 7.2 is stated to apply in the contract particulars, then the employer may assign some limited rights to a party to whom it has transferred a freehold or leasehold interest in the premises comprising the works. Among other limitations, the rights can only be assigned after practical completion. The clause does not provide a general right to assign the benefit, but does confer the right to bring proceedings in the name of the employer to enforce terms of the contract made for the benefit of the employer. It is thought that this would limit the assignee to claiming, at most, losses suffered by the employer as a result of any breach by the contractor, and would not extend to further losses suffered by itself.

Third party rights/warranties

3.50 DB16 offers two options for the granting of rights to bring a claim to persons who are not a party to the contract, either through the use of the 'third party rights' provisions included in the form or through the use of separately published standard form warranties. If either of these is to be used, full details should be set out in the 'Rights Particulars' (a separate document identified in the contract particulars (cl 7.4); Schedule 5: Part 1, paragraph 1.1).

3.51 The 'third party rights' provisions make use of the facility introduced by the Contracts (Rights of Third Parties) Act 1999. Until this Act came into force, it was a rule of English law (termed 'privity of contract') that only the two parties to a contract had the right to bring an action to enforce its terms. However, it is often the case in construction projects that other parties may wish to be in a position to take action, should one or other of the parties default on their obligations. A future owner of the property may, for example, wish to claim against the contractor should it later transpire that the project was not built according to the contract. Under the rule of privity, the future owner would be a third party, and would not be able to bring a claim. In response to this situation, 'collateral warranties' were developed which allowed third parties to pursue claims for breach of contract. Such warranties could, for example, be between contractor and owner, contractor and funder or between consultants and owner/funder.

3.52 The Contracts (Rights of Third Parties) Act 1999 changed the fundamental rules of law relating to privity by entitling third parties to enforce a right under a contract where the term in question was to provide a benefit to that third party. The third party might be specifically named, or might belong to an identified class of people. The effect of this Act is therefore to open the door to the possibility of claims being brought by a range of persons; in some cases persons that the parties to the contract may never have considered.

3.53 However, the Act allows for parties to agree that their contract will not be subject to its provisions, and many standard forms adopt this course in order to limit the parties' liability. DB16 takes this approach and under clause 1.6 states: 'Other than such rights of any Purchasers, Tenants and/or Funder as take effect pursuant to clauses 7A and/or 7B, nothing in this Contract confers or is intended to confer any right to enforce any of its terms on any person who is not a party to it.'

3.54 By invoking clause 1.6, the contract effectively excludes the terms of the Act. (In the light of the above, it is important to note that the effect of deleting or amending this clause would be significant.) The contract must define precisely which third parties will have rights with respect to the contract (in the rights particulars), and DB16 sets out what those rights will be in Schedule 5.

3.55 Schedule 5 sets out 'Third Party Rights for Purchasers and Tenants' (Part 1) and 'Third Party Rights for a Funder' (Part 2). The contractor warrants (in relation to the tenant) that it has carried out the works in accordance with the contract (with effect from practical completion), and (in relation to the funder) that it has complied with and will continue to comply with the contract. This allows both the purchaser/tenant and the funder to bring an action in respect of breaches of contract by the contractor.

3.56 There are some things to note about this system. In the case of purchasers and tenants, the contractor's liability extends to the reasonable costs of repair, renewal or reinstatement, but does not include other losses unless so stated in the rights particulars, in which case the liability will be limited to a stated maximum amount (Schedule 5: Part 1, paragraph 1.1.2). The contractor's liability is also limited in that the contract contains a net contribution clause (Schedule 5: Part 1, paragraph 1.3).[5] In addition, the contractor is entitled to rely on any term in the contract as a defence should any action be brought against it by a third party (Schedule 5: Part 1, paragraph 1.4). Where the contractor is required to take out PI insurance, it is required to provide evidence of its PI insurance to any person possessing rights under the Third Party Rights Schedule (Schedule 5: Part 1, paragraph 5). The rights may be assigned by the purchaser or tenant without the contractor's consent to another person, and by that person to a further person, but beyond this no further assignment is permitted (Schedule 5: Part 1, paragraph 6).

3.57 In the case of the funder, except for the inclusion of a net contribution clause (Schedule 5: Part 2, paragraph 1.1), no limit is placed on the extent of the contractor's liability. As above, the contractor is entitled to rely on any term in the contract, should any action be brought by the funder (Schedule 5: Part 1, paragraph 1.2), and the rights may be assigned by the funder, without the contractor's consent, to another person, and by that person to a further person, but beyond this no further assignment is permitted (Schedule 5: Part 2, paragraph 10). The Schedule also sets out various 'stepping in' rights which may be exercised by the funder in the event that it terminates its finance agreement with the employer.

3.58 Under the alternative system of 'collateral warranties' the contractor must actually enter into a warranty separately with each beneficiary. The beneficiaries are identified in the rights particulars, and the warranties are identified in clause 7C and 7D as the JCT standard forms of warranty to purchaser/tenant and funder (CWa/P&T and CWa/F). These comprise identical terms to the third party rights set out in Schedule 5.

---

[5] For guidance on the effect of such clauses, see the *CIC Risk Management Briefing: Net Contribution Clauses*, drafted by the CIC Liability Panel, chaired by the author.

Procedure with respect to third party rights and warranties

3.59　Where third party rights are to be used, the rights particulars must be drafted with care. It is important to identify the funder/purchaser/tenant, because if none is identified the rights/warranties shall not be required. It is not necessary, however, to identify a specific organisation; the description could simply be of a class of persons, e.g. 'all first purchasers' or 'the lead bank providing finance for the project' (cl 1.1). Footnote [25] to the contract particulars gives additional guidance on the rights particulars.

3.60　The third party rights take effect from the date of receipt by the contractor of the employer's notice to that effect (cl 7A.1 and 7B.1). In the case of a purchaser or tenant the notice must state their name and their interest in the works, and in the case of a funder, simply identify the party concerned. Where collateral warranties are required, the contractor must execute the stipulated warranties within 14 days of the equivalent notice from the employer (cl 7C and 7D).

3.61　Purchasers, tenants or funders are not deemed to be aware of the existence of the third party rights unless the employer gives them a copy of the relevant part of the contract. In some cases the third party may prefer to obtain a separate collateral warranty directly from the contractor, and it would be advisable for the employer to establish whether this is feasible before executing the main contract. After the contract is executed, this could only be arranged with the consent of the contractor.

3.62　If warranties are required from any sub-contractors, including named sub-contractors, this should also be set out in the rights particulars. The requirement for obtaining warranties (clause 7E) states that: 'Where the Rights Particulars state that any sub-contractor shall confer third party rights on a Purchaser, Tenant or Funder and/or to the Employer or execute and deliver a Collateral Warranty in favour of such person ... the contractor shall comply with the Contract Documents as to the obtaining of such rights or warranties'.

3.63　In the case of third party rights, the contractor is required to give notice under clause 2.26.3 or 2.26.4 of the JCT DBSub/C (cl 7E.1.1). In the case of warranties, the contractor is required, within 21 days, to take 'such steps as are required to obtain each warranty, promptly forwarding the executed document to the Employer' (cl 7E.1.2). The contractor is required to include provisions as necessary in sub-contracts in respect of the execution of required warranties (cl 3.4.2.5). The JCT publishes a standard form of sub-contract warranty (SCWa/E) to cover this situation.

4 Obligations of the contractor

4.1   The contractor's paramount obligation is to 'complete the design for the Works and carry out and complete the construction of the Works'. This obligation, which is stated in Article 1, and amplified in clause 2.1, is discussed in detail below. In addition, the contractor has important obligations in relation to progress and programming, discussed in Chapter 5, with regard to provision of information and compliance with instructions, discussed in Chapter 6, and with regard to insurance matters, discussed in Chapter 9. The contractor's obligations under DB16, which are as one would expect more extensive than under SBC16, are summarised in Table 4.1.

Table 4.1 Key obligations of the contractor

Clause	
2.1	Complete design and construct the works in compliance with the contract documents. Comply with instructions and be bound by decisions of the employer
2.1.1 and 2.1.3	Comply with all statutory instruments, etc., and give required notices; pass all approvals obtained to the employer
2.1.4	Comply with employer's instructions
2.2.3	Provide the employer with samples, etc., as stipulated in the employer's requirements or the contractor's proposals
2.2.4	Provide reasonable proof that materials and goods used comply with clause 2.2 at the employer's request
2.3	Begin the works on being given possession, proceed regularly and diligently and complete works by the completion date
2.5.1	If responsible for works insurance, notify insurers regarding use or occupation of site by the employer
2.5.2	Notify the employer of the amount of any additional premium
2.6	Permit the execution of work by persons engaged directly by the employer
2.7.3	Keep copies of contract documents and documents prepared or used for the works on site
2.7.4	Not divulge information
2.8	Provide the employer with contractor's design documents
2.10.2	Give written notice specifying divergence, if found, between employer's requirements and defined site boundary
2.13	Give written notice specifying any discrepancy or divergence, if found, between employer's requirements, the contractor's proposals and other contractor's design documents, and any instructions under the conditions

Table 4.1 Key obligations of the contractor – Continued

Clause	
2.14.1	Propose amendment to deal with clause 2.13 discrepancy within the contractor's proposals or other design documents
2.14.2	Propose amendment to deal with clause 2.13 discrepancy within the employer's requirements
2.15.1	Give notice of a divergence between the statutory requirements and the employer's requirements, the contractor's proposals or any change. Propose amendment to deal with the divergence and complete necessary design and construction work
2.16	Execute work in an emergency as necessary to comply with statutory requirements and inform the employer
2.18	Pay all statutory fees and charges, and indemnify the employer against liability in respect of them
2.19	Indemnify the employer from and against all claims and proceedings, and all damages, costs and expenses, in respect of breach of copyright, etc.
2.20.2	Notify the employer if use of documents may infringe patent rights
2.24.1	Give notice if works are delayed
2.24.2	Notify particulars of delay
2.25.6.1	Constantly use best endeavours to prevent delay
2.30	Give consent to partial possession of the works, and issue written statement
2.35	Make good defects, shrinkages and other faults specified by the employer in the schedule of defects
2.37	Provide the employer with as-built drawings and information, and maintenance/operation information
3.1	Allow the employer's agent and authorised persons access to the site and workshops
3.2	Appoint a full-time site manager, approved by the employer
3.4	Include specified provisions in any sub-contract
3.5	Comply with all instructions of the employer
3.7.1	Confirm oral instructions of the employer in writing
3.9.4	If acting as principal designer or principal contractor, notify the employer of objections to instructions effecting changes or regarding provisional sums
3.15	Use best endeavours not to disturb fossils, antiquities, etc. If found, inform the employer and take all necessary steps to preserve objects in position and condition found. Permit examination or removal of objects by a third party
3.16	Comply with CDM Regulations' requirements
3.16.2	Where acting as principal designer, prepare the health and safety file
3.16.3	Comply with regulations 8 to 10 and 13 and, where principal contractor, 12 to 14
4.8	Make applications for interim payments
4.20	Notify the employer of its initial assessment of the loss and expense likely to be incurred, and update the employer at monthly intervals
4.24.1	Submit final statement

Table 4.1 Key obligations of the contractor – Continued

Clause	
6.1	Indemnify the employer against personal injury to or death of persons
6.2	Indemnify the employer against injury or damage to property, etc.
6.4.1	Take out insurance in relation to clauses 6.1 and 6.2
6.5.1	Take out insurance against non-negligent damage to property
6.9	Ensure Schedule 3 Insurance Option A.1 and clause 6.9 joint names insurance policy covers sub-contractors
6.10.1	Take out terrorism cover as extension to insurance policy (Insurance Option A)
6.11.1	Give notice if availability of terrorism cover ceases
6.12	Provide documentary evidence of insurance cover
6.13.1	Give notice to the employer of damage to work or materials
6.13.3	Authorise payment of insurance monies to the employer
6.13.4	Restore damage to work or materials
6.15	Take out PI insurance
6.16	Give notice if PI insurance not available at reasonable rates
6.18	Comply with the Joint Fire Code
6.19.1	Ensure remedial measures are carried out in accordance with a notice of remedial measures
8.5.2	Inform the employer of an insolvency event
8.5.3.3	Allow the employer to take measures following an insolvency event
8.7.2.1	Remove tools, etc. from the site
8.7.2.2	In the event of termination through insolvency by the contractor, provide the employer with copies of the contractor's design documents
8.7.2.3	Assign benefits of sub-contracts to the employer
8.12.2.1	Remove tools, etc. from the site
8.12.3	Prepare account after termination
8.12.3	Provide documents for preparation of account after termination
Schedule 1:	
1	Submit drawings, etc. as set out in employer's requirements
Schedule 2:	
1.1.1	Enter into contracts with named sub-contractors and notify the employer of dates
1.1.2	Notify the employer of reasons when the contractor is unable to enter into a sub-contract
1.3.2	Obtain consent of the employer before terminating a named sub-contractor's employment
1.4.1	Complete named sub-contract work
2.2	Submit an estimate of the valuation of a change
2.4	Take all reasonable steps to agree estimate with the employer
3.2	Submit an estimate of the amount of direct loss and/or expense incurred

Table 4.1 Key obligations of the contractor – Continued	
Clause	
4.1.1	Provide an acceleration quotation if invited by the employer
5	Work in a collaborative manner with the employer and other team members
6.1	Endeavour to maintain a working environment in which health and safety is of paramount concern
6.2	Comply with health and safety codes of practice, etc.
7.2	Provide the employer with details of proposed cost-saving changes
8.2	Provide the employer with requested information on the environmental impact of contractor-selected materials
9.2	Provide the employer with information to enable the contractor's performance to be monitored against performance indicators
10	Notify the employer promptly of any matter likely to give rise to a dispute and meet to negotiate in good faith to resolve the matter
12.2	Include particular provisions in sub-contracts
Schedule 3:	
Insurance Option A – New Buildings – All-Risks Insurance of the Works by the Contractor	
1	Take out a joint names insurance policy for the works

4.2 Should Supplemental Provision 5 be incorporated, the contractor would, in addition to these primary obligations, be under an express duty of collaboration. The provision states:

> The Parties shall work with each other and with other project team members in a co-operative and collaborative manner, in good faith and in a spirit of trust and respect. To that end, each shall support collaborative behaviour and address behaviour which is not collaborative.

This places a duty on the contractor to collaborate not only with the employer, but also with other team members, which would include the employer's appointed consultants. Other supplemental provisions introduce further obligations, including to notify the employer promptly of any matter that may give rise to a dispute. Such obligations may affect the interpretation of the nature and extent of the contractor's duties under other clauses.

The design obligation

4.3 The contractor is required to complete the design in accordance with the employer's requirements and under clause 2.1 this obligation also extends to compliance with any change to the requirements instructed by the employer. Clause 2.11 makes it clear that the contractor is not responsible for the contents of the employer's requirements, or for verifying the adequacy of any design contained within them. This clause is included to prevent such an obligation being implied, as it was in the case of *Co-operative Insurance Society* v *Henry Boot*. Although it is not entirely clear, it is unlikely to prevent the implication of a 'duty to warn' regarding any other aspects of the consultant team's design; for example, where the design is varied through an instruction (for an instance of this see the earlier case of *Plant Construction* v *Clive Adams*). A certain degree of vigilance should be expected from the contractor, particularly where there is an express duty of collaboration as described above.

> *Co-operative Insurance Society* v *Henry Boot Scotland and others* (2002) 84 Con LR 164
>
> The Co-operative Insurance Society (the Society) engaged the contractor Henry Boot on an amended version of JCT80 incorporating the Designed Portion Supplement, where the relevant terms are virtually identical to those of WCD98. During construction, problems arose where soil and water flooded into a basement excavation. An engineer had originally been employed by the Society to prepare a concept design for the structure, and Henry Boot had developed the design and prepared working drawings. The Society brought claims against Henry Boot and the engineers. Henry Boot argued that their liability was limited to the preparation of the working drawings. The judge, however, took the view that completing the design of the contiguous bored pile walls included examining the design at the point when it was taken over, assessing the assumptions on which it was based and forming a view as to whether they were appropriate.

> *Plant Construction* v *Clive Adams Associates and JMH Construction Services* [2000] BLR 137 (CA)
>
> Ford Motor Company engaged Plant on a JCT WCD contract to design and construct two pits for engine mount rigs at Ford's research and engineering centre in Essex. Part of the work included underpinning an existing column and in the course of the work temporary support was required to the column and the floor above. JMH was sub-contracted to carry out this concrete work. Ford's own engineer gave instructions regarding the temporary supports, which comprised four Acrow props. JMH and Plant's engineers, Clive Adams Associates, felt the props to be inadequate and discussed this on site. The support was installed as instructed and failed, so that a large part of a concrete floor slab collapsed. Plant settled with Ford, and brought a claim against JMH and Clive Adams (who settled). The court found that the duties of the sub-contractor included warning of any aspect of the design that it knew to be unsafe. It reserved its opinion on whether the duty would extend to unsafe aspects it ought to have known, or design errors that were not unsafe.

4.4 As discussed above, the third recital states that 'the Employer has examined the Contractor's Proposals and, subject to the Conditions, is satisfied that they appear to meet the Employer's Requirements', which implies that, insofar as the design has been finalised at the time of acceptance of tender, the employer has accepted the solution. However, there are convincing arguments that, despite this recital, the contractor remains obliged to meet all the employer's requirements, regardless of anything set out in the initial contractor's proposals (see paragraph 3.42), and therefore should continue to monitor the developing design and adjust the solution as soon as any discrepancy with the employer's requirements is discovered.

4.5 It is important to note that under clause 2.17.1 the contractor's liability is stated to be equivalent to that of 'an architect or other appropriate professional designer'. In effect, this statement reduces the strict obligation that would normally have been implied to a requirement to use due skill and care (see paragraph 1.31). It would be clearer if the form simply stated this as, strictly speaking, an architect's liability will depend on the terms of engagement used in each case, but the phrase is intended to mean the liability that would normally be implied by law. In practical terms, the employer will not be able to claim against the contractor for every defect or problem inherent in the design, but only when the employer can show that the design was not prepared with the skill and care to be expected of a competent architect undertaking such work.

4.6   The employer's requirements may state that any selection of materials should be fit for the purposes set out in the employer's requirements, but this should not be confused with what is commonly referred to as a 'fitness for purpose' obligation, as by clause 2.17.1 the contractor's liability is limited to the use of reasonable skill and care in making that selection.

4.7   Clause 2.17.2 states that, where the contract involves work in connection with a dwelling, 'the clause 2.17.1 reference to the Contractor's liability includes liability under the Defective Premises Act 1972'. The effect is that the contractor is liable to the employer for design in relation to a dwelling to the same extent that an architect would be liable. The duty is expressed as 'to see that the work which he takes on is done in a workmanlike or, as the case may be, professional manner, with proper materials and so that as regards that work the dwelling will be fit for habitation when completed' (section 1(1)). In case law, the duty has not generally been taken to be a strict or absolute warranty of fitness (*Alexander and another* v *Mercouris*), although other authorities suggest that it is a strict duty. It should be noted that, although the contractor's liability is limited to the amount stated in the contract particulars, the limitation does not apply to work in connection with a dwelling (cl 2.17.3).

> *Alexander and another v Mercouris* [1979] 1 WLR 1270
>
> This case considered when the duty arose, not its scope, but some observations are helpful, for example Lord Justice Buckley stated: 'It seems to me clear upon the language of Section 1(1) that the duty is intended to arise when a person takes on the work. The word "owes" is used in the present tense and the duty is not to ensure that the work has been done in a proper and workmanlike manner with proper materials so that the dwelling is fit for habitation when completed, but to see that the work is done in a proper and workmanlike manner with proper materials so that the work will be fit for habitation when completed. The duty is one to be performed during the carrying on of the work. The reference to the dwelling being fit for habitation indicates the intended consequence of the proper performance of the duty and provides a measure of the standard of the requisite work and materials. It is not, I think, part of the duty itself.'

4.8   The employer has an option to limit the contractor's liability for the consequential losses arising from its failure to meet the design obligations, in addition to any liability for liquidated damages. This limitation is, in fact, less extensive than it might seem, as it protects the contractor only from claims of special types of damage, and not from the losses that would normally be foreseeable as a direct result of a breach of the contractor's design obligations. Nevertheless, if the employer does agree to this limitation, this should be reflected in a reduced tender price. The limitation of contractor's liability is set up under the contract particulars in clause 2.17.3. If no sum is inserted into the contract particulars, the contractor's liability in respect of these matters will be unlimited.

4.9   If the employer wishes to impose a higher level of liability, this would require amendments to the form, which should be set out in the employer's requirements. The employer would be unwise to make any such amendments without taking legal advice. It is interesting to note that both the RIBA and NEC3 forms contain optional 'fitness for purpose' liability clauses, which could serve as a useful starting point were such amendments required.

4.10 The contractor is required to hold PI insurance to cover its liability under clause 6.15.1 of the type and in an amount not less than stated in the contract particulars. Advice should be taken on the type and extent of insurance required, which would normally be in the form of a PI policy. Insurance to cover any stricter liability is not usually required, and although it would be possible to obtain, such insurance might be expensive.

Standards and quality

4.11 Under clause 2.1, where the employer's requirements specify particular materials, goods or workmanship, the contractor must provide these items in accordance with the requirements, and this obligation is amplified in clause 2.2. The contractor will be relieved of any obligation to provide materials, goods or workmanship fit for the intended use, and its obligation will be limited to supplying those items specified. It would be implied, of course, that the items should be of merchantable quality. If they are not described in the requirements, the contractor must provide materials and goods in accordance with the proposals, or with 'other Contractor's Design Documents', i.e. documents prepared or used by the contractor in connection with the works.

4.12 If the contractor wishes to substitute any materials or goods for those specified, it must obtain the permission of the employer in writing, which may not be unreasonably delayed or withheld (cl 2.2.1 and 1.10). The contract does not set out what should happen when the employer has given such consent. If the material is of the same standard and value, then the employer would have no reason for concern, but if the substitution were of less value, the employer would reasonably wish to see a reduction in the amount to be paid. If the material was originally specified in the employer's requirements, the employer could issue an instruction requiring a change, which would be valued in the normal way. If, however, the material was specified in the proposals or further specifications submitted to the employer, then it is not clear how this adjustment might be handled under the contract. The employer could again issue a change instruction, specifying the material proposed as a new employer's requirement, but this is a somewhat artificial method of incorporating what is, in effect, a change to the contractor's proposals. An option would be to agree a reduction, which should be recorded in writing and signed by both parties.

4.13 As in SBC16, the obligation to provide materials in accordance with the contract is qualified by the phrase 'so far as procurable'. This means that failure to supply materials would not be a breach of contract by the contractor were the item genuinely unavailable. However, 'procurable' would be interpreted on the assumption that reasonable steps had been taken to order the item in advance, and would not excuse a contractor who had merely forgotten to place an order. How far in advance such orders should be placed will depend on the particular circumstances and, if the parties cannot reach agreement, the matter could be referred to adjudication.

4.14 The contract does not state what should happen if a material is not procurable. It is suggested that, as part of its overall design obligation, the onus would be on the contractor to put forward a suggestion for the employer's approval. Any necessary adjustment to the contract sum would be dealt with as discussed under paragraph 4.12 above.

4.15 If the employer's requirements state that any items should be 'to approval', this means that the contractor only fulfils its obligations in respect of quality if the employer is satisfied. It is very important to note, however, that the final statement (or employer's final statement) is conclusive evidence that, where the employer's requirements or any change have expressly stated that the quality is to be to the approval of the employer, then the employer is so satisfied (cl 1.8.1.1). This would have the effect of preventing the employer from bringing a claim regarding those items of work. It would therefore be advisable for the employer to avoid using phrases such as 'to approval' or 'to the employer's satisfaction' in the requirements or in any change instruction. The dangers of including them were underlined in the case of *London Borough of Barking & Dagenham* v *Terrapin Construction Ltd*.

> *London Borough of Barking & Dagenham* v *Terrapin Construction Ltd* [2000] BLR 479
>
> The Borough employed Terrapin Construction to design and build new and refurbishment work at a school in Dagenham. No document entitled 'Employer's Requirements' had been issued to the contractor at tender stage, but the contractor had been given a 'brief', which set out in general terms the nature of the works which the Borough wanted to have designed, and the court decided that requirements were 'represented by the contract as a whole'. The contract was to be on WCD81. Once a tender figure had been negotiated, the Borough sent the contractor an order for the work, which set out the agreed contract figure, incorporated the terms of WCD81 and stated: 'In consideration of this Agreement hereinafter contained on the part of the employer the Contractor shall and will execute complete and maintain the works in all respects to the satisfaction of the Controller of Development and Technical Services'. The court decided that in this context the final statement was conclusive evidence that all work had been carried out to the satisfaction of the employer.

4.16 However, if the phrase 'or otherwise approved' is used in a specification or bill of quantities, this means neither that the employer must be prepared to consider alternatives put forward by the contractor, nor that the employer must give any reasons for rejecting alternatives (*Leedsford* v *City of Bradford*). It merely gives the employer the right to do so. It should be noted that under clause 2.2.3 the employer may require the contractor to provide samples. This is discussed under paragraph 6.11.

> *Leedsford Ltd* v *The Lord Mayor, Alderman and Citizens of the City of Bradford* (1956) 24 BLR 45 (CA)
>
> In a contract for the provision of a new infant school, the contract bills stated 'Artificial Stone ... The following to be obtained from the Empire Stone Company Limited, 326 Deansgate, or other approved firm'. During the course of the contract the contractor obtained quotes from other companies and sent them to the architect for approval. The architect, however, insisted that Empire Stone was used and, as Empire Stone was considerably more expensive, the contractor brought a claim for damages for breach of contract. The court dismissed the claim stating: 'The builder agrees to supply artificial stone. The stone has to be Empire Stone unless the parties agree some other stone, and no other stone can be substituted except by mutual agreement. The builder fulfils his contract if he provides Empire Stone, whether the Bradford Corporation want it or not; and the Corporation Architect can say that he will approve of no other stone except the Empire Stone' (Hodson LJ at page 58).

Obligations in respect of quality of sub-contracted work

4.17 Where work is sub-contracted, the contract is clear that the contractor will still have ultimate responsibility for the standard of workmanship, materials and goods provided by the sub-contractor (cl 3.3.1). This obligation extends to any sub-let design work, and is the same whether or not the sub-contractor was named in the employer's requirements.

Compliance with statute

4.18   The contractor is under a statutory duty to comply with all legislation relevant to the carrying out of the works, for example in respect of goods and services, building and construction regulations and health and safety. The duty is absolute and it is not possible to contract out of any of the resulting obligations.

4.19   DB16 includes a contractual duty in addition to the statutory duty, which gives additional protection to the employer, in that failure to comply with statute becomes a breach of contract. Under clause 2.1.1 the contractor is obliged to comply with all statutory requirements except where the employer's requirements specifically state that the employer's requirements (or relevant parts) comply (cl 2.1.2). In emergencies, the contractor may carry out work in order to achieve compliance with statutory requirements without the prior consent of the employer, but must notify the employer forthwith (cl 2.16).

4.20   Clause 2.1.1 requires the contractor to complete the works in compliance with statutory requirements and 'for that purpose shall complete the design for the Works … and shall give all notices required by the Statutory Requirements'. This obligation therefore extends to obtaining statutory permissions as relevant, unless the employer's requirements state that they comply, and would include obtaining planning permission. It also includes complying with Building Regulations, and the contractor should make sufficient allowance in its tender to cover all permissions, testing, commissioning and certification required. All consents and permissions must be passed to the employer (cl 2.1.3).

4.21   The duty of statutory compliance applies even when aspects of the employer's requirements do not in fact comply with a statutory requirement. The contractor is relieved of this duty only when the requirements make a positive assertion of compliance.

4.22   The contractor must pay any fees or charges and is reimbursed for these only where the tender documents included a provisional sum for this item (cl 2.18). Clause 2.18 also requires the contractor to indemnify the employer against any liability in respect of such fees and charges. Should the contractor fail to pay any fee, the employer would be able to claim any losses resulting from such failure from the contractor, including, for example, any fines or legal fees incurred by the employer.

4.23   If either the contractor or the employer finds any divergence between the employer's requirements or the contractor's proposals or other contractor's design documents and statutory requirements, then one must give immediate written notice to the other (cl 2.15.1). The contractor must then submit a proposal for dealing with the problem and, with the employer's consent, give effect to the amendment entirely at its own cost. The employer is required to make a note of the amendment on the contract documents. It is not clear why the JCT has included this requirement, as it does not apply to any other type of discrepancy or divergence. It is suggested that if the employer dislikes the contractor's proposal for dealing with the divergence, it must issue an instruction effecting an appropriate change to the requirements. The only cases where the contractor could claim additional costs would be where the statutory requirement did not exist at the base date, where it became necessary to conform with conditions attached to an approval obtained after the base date, or where the employer's requirements asserted that the item in question complied (cl 2.15.2). In all these cases the adjustment is to be treated as if it were a change in the employer's requirements.

4.24 The contractor is under no express obligation to search for any divergence, but if it fails to find a divergence and the resulting works do not comply, then the contractor will be in breach of clause 2.1.2 notwithstanding that the error lay in the employer's requirements.

4.25 Under clause 2.15.1, it is clear that the contractor remains generally responsible for any non-compliance of the requirements or the proposals with statutory requirements which exist at the base date. Nevertheless, although this obligation is extensive, it should be noted that delays to progress of the works, which the contractor has taken all reasonable steps to avoid, that are due to failure to obtain approval from statutory authorities in good time are grounds for an extension of time (cl 2.26.13). Additionally, delays relating to planning permission are also grounds for claiming direct loss and/or expense (cl 4.21.4). Under clause 8.11.1.6, such delays may be grounds for termination where they lead to a suspension of the works.

4.26 Under clause 2.15.2.1, if there is any change in statutory requirements after the base date which results in a change to the contractor's proposals, this is treated as if it were a change, which means that it would fall to be valued under clause 5.1, and may give rise to a claim for an extension of time under clause 2.26.1, or for loss and/or expense under clause 4.21.1, and may be grounds for termination under clause 8.11.1.2. Similarly, under clause 2.15.2.2, if the contractor's proposals are affected by 'the terms of any permission or approval made by a decision of the relevant authority after the Base Date' this should also be treated as a change. This could include decisions relating to an approval obtained after the base date, for example planning permission obtained by the contractor, or where an approval had previously been obtained by the employer but contained reserved matters on which an authority later made a decision. The clause 2.15.2.2 obligation is subject to the proviso that the employer's requirements have not precluded such adjustments from being treated as a change. In other words, the employer can elect that decisions of authorities are a matter for the contractor to absorb in its original price. Requiring the contractor to take such a risk is, however, likely to result in an increase in the tender figure.

4.27 If an amendment becomes necessary because it is found that a part of the employer's requirements which specifically states that it complies with statutory requirements does not, in fact, comply, then in such a case the employer must issue an instruction under clause 3.9 effecting a change (cl 2.15.2.3).

Health and safety legislation

4.28 The contract allows for the contractor to take on the roles of principal designer and principal contractor under the CDM Regulations 2015; in fact, under Articles 5 and 6 the contractor will be responsible for those roles unless another name is inserted. The Regulations define the principal designer as 'a designer with control over the pre-construction phase' (regulation 5.1(a)). It is unlikely that the contractor would be in a position to take on this role unless a separate appointment is made at a very early stage in the project, before the DB16 contract is entered into. However, the contractor will, under the terms of DB16, be responsible for the developing design after the contract is formed, and therefore it would be sensible for the contractor to act as principal designer from this point on. It would be possible for the employer to replace the original principal designer with the contractor once the tender negotiations are concluded, provided, of course, that the contractor is sufficiently competent and has the resources to take on this role (which it ought to be if offering design services). If appointed as principal designer, the contractor is required under clause 3.16.2 to comply with all the relevant duties under the CDM Regulations.

4.29 Where the contractor is not the principal designer, clause 3.16.1 places a contractual obligation on the employer to ensure that the principal designer carries out his or her duties under the CDM Regulations. There are equivalent provisions where the contractor is not the principal contractor. This is a wider obligation than the 'reasonable satisfaction with competence' obligation imposed by the Regulations. Breach of this clause gives the contractor the right to terminate the contract under clause 8.9.1.3. It is more likely, in practice, that the contractor will claim for an extension of time or direct loss and/or expense for breach of clause 3.16, as this is a relevant event under clause 2.26.6 and a relevant matter under clause 4.21.5. An example might be where a principal designer delays in commenting on a contractor's proposed amendment to the construction phase plan, and progress is thereby delayed.

4.30 Clause 3.16.3 places a duty on the contractor, if acting as the principal contractor, to comply with all the relevant duties set out in the CDM Regulations 2015. Breach of this duty gives grounds for termination under clause 8.4.1.5. The warning notice has still to be given, and JCT Practice Note 27 suggested that the provision should be used only for situations where the Health and Safety Executive is likely to close the site. Any breach is covered, however, provided termination is not unreasonable or vexatious, and the employer should consider any breach that might lead to action being taken against the employer as a serious one. If work needs to be postponed or other instructions given, due to a breach by the contractor, then there should be no entitlement to an extension of time or a claim for direct loss and/or expense.

4.31 The contractor should take the cost of compliance with the CDM Regulations (e.g. the cost of developing the construction phase plan) into account at the tender stage. No claims may be made for compliance with the Regulations (e.g. adjusting the construction phase plan to suit the contractor's or sub-contractor's working methods) and no extension of time will be given (cl 3.16.4). If alterations are needed as a result of an instruction requiring a change, then the costs are included in valuing the change and the alterations may be taken into account in assessing an application for an extension of time.

4.32 The CDM Regulations are not the only statutory health and safety obligations which may apply to the project, and the contractor must comply with all applicable health and safety laws. Supplemental Provision 6, furthermore, states that 'the Parties will endeavour to establish and maintain a culture and working environment in which health and safety is of paramount concern', which suggests that a 'best practice' rather than a minimum compliance approach is required. The provision sets out several specific requirements: for example, the contractor undertakes to comply with all approved codes of practice, to ensure that personnel receive induction training and have access to advice and to ensure that there is full and proper health and safety consultation with all such personnel in accordance with the Health and Safety (Consultation with Employees) Regulations 1996.

5 Possession and completion

5.1 The date of possession or commencement and the date for completion of the works are key dates in any building contract, and DB16 requires a 'Date of Possession' and a 'Date for Completion' to be inserted into the contract particulars. DB16 also offers the facility for the work to be carried out in phases. If phased completion is required, then the work must be split into clearly identified sections, and a separate date of possession and date for completion entered for each section. The contractor is required to take possession on the date of possession and complete by the date for completion, and failure to do so may give rise to a liability for liquidated damages. Provisions exist for deferring the date of possession, and extending the completion date. If the work is divided into sections, provisions for commencement, completion, deferment and extension operate independently for each section.

5.2 Although the employer may state preferred dates at the time of tendering, it is likely that in design and build procurement the dates will be subject to negotiation. Nevertheless, these should be finalised before the contract is entered into and before the contractor commences work. In the event that work is started before dates are finally agreed, the contract will be subject to the Supply of Goods and Services Act 1982 or the Consumer Rights Act 2015, which state that the work is to be completed within a reasonable time. However, once the job is under way it is unlikely that the parties will be able to agree on a definition of 'reasonable time'.

Possession by the contractor

5.3 Possession of the site is a fundamental term of the contract. If the employer fails to give the contractor possession of the site on the agreed date this would be a breach of contract, which would give the contractor the right to claim an extension of time under clause 2.26.6 and direct loss and/or expense under clause 4.21.5. Any extended failure would be considered a serious breach by the employer, which might amount to repudiation, and therefore give the contractor the right to treat the contract as being at an end. Giving possession of part of the site only, or giving possession in stages, could amount to a breach unless this intention has been made clear in the contract documents (*Whittal Builders* v *Chester-le-Street DC*).

> ***Whittal Builders Co. Ltd*** v ***Chester-le-Street District Council*** (1987) 40 BLR 82
>
> Whittal Builders contracted with the Council on JCT63 to carry out modernisation work to 90 dwellings. The contract documents did not mention the possibility of phasing, but the Council gave the contractor possession of the houses in a piecemeal manner. Even though work of this nature is frequently phased, the judge nevertheless found that the employer was in breach of contract for not giving the contractor possession of all 90 dwellings at the start of the contract and the contractor was entitled to damages.

5.4 Possession of the site must be given in such a way that the contractor is not prevented from working in whatever way or sequence it chooses. With most jobs, this means that the contractor must be given clear possession of the whole site until practical completion. Clause 2.3 reinforces this by stating that 'the Employer shall not be entitled to take possession of any part or parts of the Works' until the date of issue of the practical completion statement. Where clear possession is not intended, then the tender documents should set out the restrictions in detail and the contract must be amended accordingly. This would include any employer restrictions relating to access to the site. Should the employer wish to use any part of the works for any purpose during the time that the contractor has possession, this should also be made clear in the tender documents, otherwise such use can only be with the agreement of the contractor (cl 2.5.1).

5.5 If clause 2.4 has been stated in the contract particulars to apply, then it is possible for the employer to defer possession without the agreement of the contractor. The clause allows for deferment for a period not exceeding six weeks, and the maximum period required must be stated in the tender documents and inserted into the contract particulars (cl 2.4). Any delay beyond the period stated in the contract particulars is a breach of contract. Although the contract does not specifically require it, the employer should give written notice of its intention to defer as far in advance of the commencement date as is practicable. If possession is deferred, the contractor may claim an extension of time (cl 2.26.3) and direct loss and/or expense (cl 4.19.1).

5.6 Under clause 3.10, the employer may issue instructions regarding postponement of any design or construction work. If work is postponed, the contractor may claim an extension of time (cl 2.26.2) and direct loss and/or expense (cl 4.21.2).

5.7 The parties are always free to renegotiate the terms of any contract. Therefore, if a delay in giving possession arises which is longer than the period stated in the contract particulars, the parties would have to agree a new date of possession, usually involving financial compensation for the contractor. Any further delay beyond the agreed date would, of course, be a breach.

Progress

5.8 It would normally be implied within a construction contract that a contractor will proceed 'regularly and diligently', and this is an express term in DB16 (cl 2.3). The contractor is required to commence work on being given possession of the site, to proceed regularly and diligently and to complete the works on or before the completion date. The contractor is free to organise its own working methods and sequences of operations, with the qualification that it must comply with statutory requirements and the construction phase plan. This has been held to be the case even where the contractor's chosen sequencing may cause extra cost to the employer due to the operation of fluctuations provisions (*GLC* v *Cleveland Bridge and Engineering*).

> *Greater London Council* v *Cleveland Bridge and Engineering Co.* (1986) 34 BLR 50 (CA)
>
> The Greater London Council (GLC) employed Cleveland Bridge to fabricate and install gates and gate arms for the Thames Barrier. The specially drafted contract provided dates by which Cleveland Bridge had to complete certain parts of the works. Clause 51 was a fluctuations

> provision which allowed for adjustments to be made to the contract sum if, for example, the rates of wages or prices of materials rose or fell during the course of the contract. The clause also contained the phrase 'provided that no account shall be taken of any amount by which any cost incurred by the Contractor has been increased by the default or negligence of the Contractor'. The contract was lengthy, and Cleveland Bridge left part of the work to be carried out at the very end of the period, but delivered the gates on time. The result was that the GLC had to pay a large amount of fluctuations in respect of the work done at the last minute. The GLC argued that the contractor had failed to proceed regularly and diligently, and therefore was in default. The court held that even if the slowness of the contractor's progress might at certain points have given the employer the right to terminate the contract under the termination provisions, this would not, in itself, be a breach of contract as referred to in clause 51. The contractor could organise the work in any way it wished, provided it completed on time: it was therefore owed the full amount of the fluctuations.

5.9 There is no requirement under DB16 for the contractor to produce a programme. Of course, there is nothing to prevent such a requirement being included in the employer's requirements, but it should be made clear that this will not be a contract document. If a programme is required, it is advisable to provide that the contractor submits it as part of the proposals, or at least before the contract is entered into. It is common practice to ask for a detailed programme indicating activities, a critical path and resources, and possibly key dates when information regarding the developing design will be supplied to the employer.

5.10 It is notoriously difficult to extract programmes from contractors once work has commenced. Sometimes the programme, when finally produced, can include an element of post-rationalisation to show that early events caused problems and delays. It is also advisable to ask for a critical path analysis, since it can otherwise be difficult to assess the effect of delays.

5.11 As the employer is not required to issue any further information, the programme does not fulfil the same role as under SBC16 in terms of identifying when such information may be needed. However, it will be an important tool in alerting the employer as to the latest date by which any final decisions should be made, or any changes could be made without causing delays.

Completion

5.12 The contractor is obliged to complete the works by the completion date. The most important reason for giving an exact completion date in a building contract is that it provides a fixed point from which damages may be payable in the event of non-completion. Generally, in construction contracts, the damages are 'liquidated', and typically are fixed at a rate per week of overrun. The contractor may finish the works earlier, and the employer would be obliged to accept the building, even if it were not convenient to do so.

5.13 The contractor, in general, accepts the risk of all events that might prevent completion by this date, except to the extent that the contract provides otherwise. The contractor would normally be relieved of this obligation if the employer caused delays or in some way prevented completion. In order to avoid situations arising where the contractor is no longer bound by the completion date, most contracts contain provisions allowing for the adjustment of the completion date in the event of certain delays caused by the employer.

5.14   In contracts it is sometimes essential that completion is achieved by a particular date and failure to meet that date would render the result worthless. This is sometimes referred to as 'time is of the essence'. Breach of such a term would be considered a fundamental breach, and would give the employer the right to terminate the contract and treat all its own obligations as at an end. The expression 'time is of the essence' is seldom, if ever, applicable to building contracts such as DB16, as the inclusion of extensions of time and liquidated damages provisions imply that the parties intended otherwise.

5.15   In DB16 a 'Date for Completion' is inserted into the contract particulars (or where the sectional completion option is used, a separate date will be stated for each section). The contractor is obliged to complete by the 'Completion Date' (cl 2.3), which is defined as the 'Date for Completion of the Works or of a Section … or such other date as is fixed either under clause 2.25 or by a Pre-agreed Adjustment' (clause 2.25 contains the DB16 provisions for the granting of extensions of time). If the contractor fails to complete by the completion date, liquidated damages become payable (see Figure 5.1).

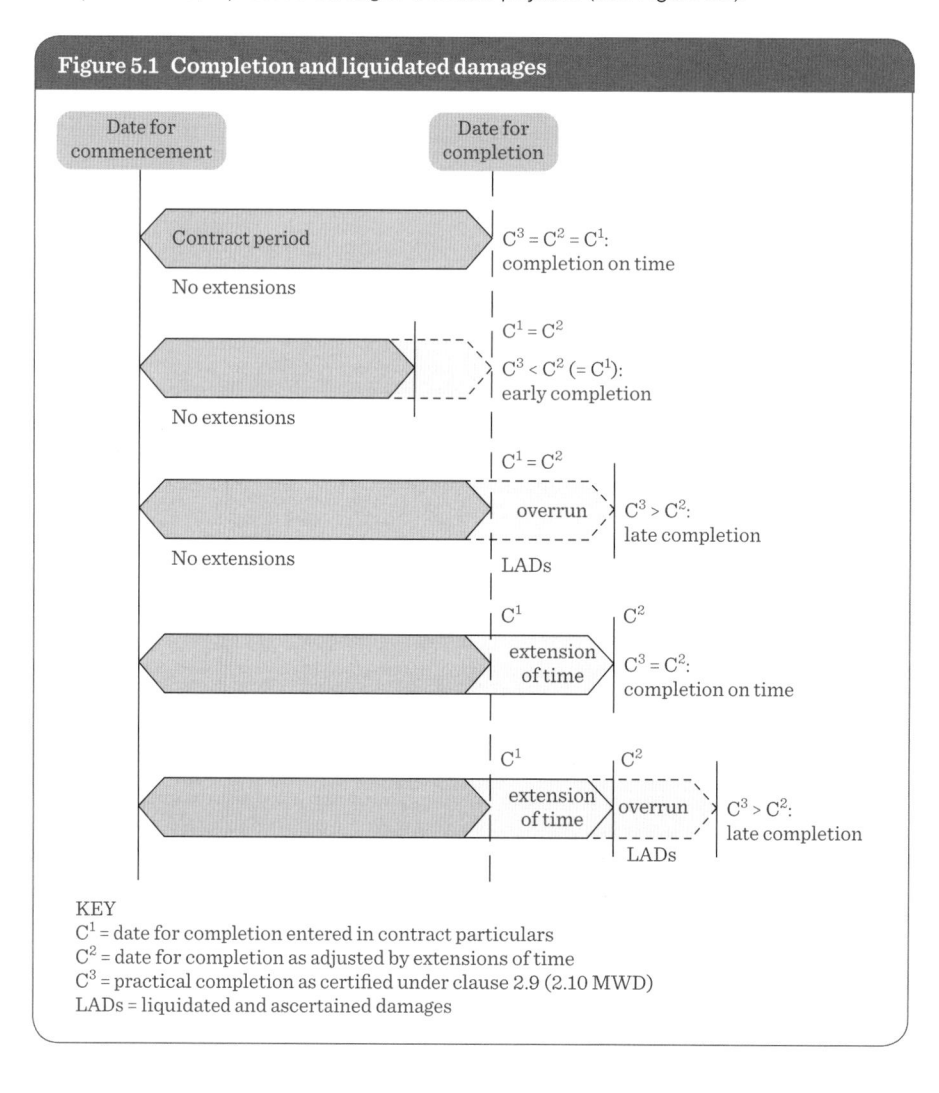

Figure 5.1 Completion and liquidated damages

KEY
C^1 = date for completion entered in contract particulars
C^2 = date for completion as adjusted by extensions of time
C^3 = practical completion as certified under clause 2.9 (2.10 MWD)
LADs = liquidated and ascertained damages

Pre-agreed adjustment

5.16 There are two processes that can result in a 'Pre-agreed Adjustment' to the completion date (cl 2.23.2). The first process is where the employer operates the 'Valuation of Change' mechanism set out in paragraph 2 of the supplemental provisions. Under Schedule 2 paragraph 2.3, the contractor is required to estimate not only the addition to the contract sum (see paragraph 7.7), but also any extension of time appropriate for the proposed changes. If the employer and contractor can subsequently agree on an amount, then this is binding on the parties.

5.17 The second process is through the confirmed acceptance of an acceleration quotation (paragraph 4 of the Schedule 2 supplemental provisions). This provision allows the employer to investigate the possibility of achieving practical completion before the completion date. If invited by the employer, the contractor must provide, within 21 days, a quotation identifying the time that can be saved and the required increase in the contract sum, or explain why any acceleration is impracticable (paragraph 4.1.1). The employer must accept or reject the proposal within seven days, and a confirmed acceptance would be binding on the parties. If the employer does not accept the quotation, the contractor is paid a reasonable sum for the cost of its preparation (paragraph 4.4.1).

Extensions of time

Principle

5.18 As there is no independent contract administrator in DB16, it is important that the employer operates the extension of time provisions fairly, and understands the reason for their inclusion. They are not included as a let-out clause for the contractor, relieving it of the obligation to finish by the agreed date, and deleting these provisions would not benefit the employer; in fact, it would cause serious problems. If no such provisions were included, and a delay occurred that was caused by the employer, a court would consider that the contractor was no longer bound to complete by the completion date (*Peak Construction v McKinney Foundations*). The employer would therefore lose the right to liquidated damages, even though some of the blame for the delay might rest with the contractor. The extension of time provisions are therefore included to preserve the employer's right to liquidated damages, in the event that the contractor fails to complete on time, due in part to some action for which the employer is responsible. The phrase 'time at large' is often used to describe this situation. However, this is, strictly speaking, a misuse of the phrase as in most cases the contractor would remain under an obligation to complete within a reasonable time.

> *Peak Construction (Liverpool) Ltd v McKinney Foundations Ltd* (1970) 1 BLR 111 (CA)
>
> Peak Construction was the main contractor on a project to construct a multi-storey block of flats for Liverpool Corporation. The main contract was not on any of the standard forms, but was drawn up by the Corporation. McKinney Foundations Ltd was the sub-contractor nominated to design and construct the piling. After the piling was complete and the sub-contractor had left the site, serious defects were discovered in one of the piles and, following further investigation, minor defects were found in several other piles. Work was halted while the best strategy for remedial work was debated between the parties. The city surveyor did not accept

> the initial remedial proposals, and it was agreed that an independent engineer would prepare an alternative proposal. The Corporation refused to agree to accept his decision in advance, and delayed making the appointment. Altogether it was 58 weeks before work resumed (although the remedial work took only six weeks) and the main contractors brought a claim against the sub-contractor for damages. The Official Referee, at first instance, found that the entire 58 weeks constituted delay caused by the nominated sub-contractor and awarded £40,000 damages for breach of contract, based in part on liquidated damages which the Corporation had claimed from the contractor. McKinney appealed, and the Court of Appeal found that the 58-week delay could not possibly entirely be due to the sub-contractor's breach, but was in part caused by the tardiness of the Corporation. This being the case, and as there were no provisions in the contract for extending time for delay on the part of the local authority, it lost its right to claim liquidated damages, and this component of the damages awarded against the sub-contractor was disallowed. Even if the contract had contained such a provision, the failure of the architect to exercise it would have prevented the Corporation from claiming liquidated damages. The only remedy would have been for the Corporation to prove what damages it had suffered as a result of the breach.

5.19 DB16 therefore includes provisions for extending the time for completion where the employer has caused delays. It also allows for extensions of time in the case of certain neutral events, such as exceptionally adverse weather. In contrast to employer-caused delays, it would be possible to require the contractor to accept the risk of these neutral events, and consequently pay losses to the employer should it fail to adjust its programme to compensate for these risks. Taking on this risk would no doubt result in a higher tender figure. In some projects the employer may be prepared to accept this increase in price in exchange for greater certainty regarding completion. This would require careful amendment of the form, which should not be undertaken without legal advice.

Procedure

5.20 In DB16 the provisions for granting an extension of time (other than pre-agreed adjustments) are contained in clause 2.25. Where the sectional completion option is used, the provisions apply independently to each section. As noted above, it is particularly important that the employer takes great care to operate these provisions fairly, and neither delays the award of extensions, nor takes an overly conservative approach to their assessment. Any default by the employer is likely to result in the contractor asserting that it is no longer bound to finish by the completion date.

5.21 The contractor must give written notice 'forthwith' to the employer, when it appears that progress is being or is likely to be delayed (cl 2.24.1). The notice must be given whether or not completion is likely to be delayed, and irrespective of the cause (i.e. the requirement to give notice is not limited to circumstances where the contractor is claiming an extension of time). The use of the terms 'forthwith' and 'is likely' suggests that the notice should be given as soon as a potential problem becomes apparent, without waiting to see whether it results in a measurable delay (this requirement could be reinforced by the adoption of Supplemental Provision 10). It appears that, prior to practical completion, notification by the contractor is a condition precedent to the award of an extension of time; in other words, the employer may not grant an extension unless a valid notice has been given (although this is not the case in relation to extensions awarded after practical completion, see paragraph 5.28 below). The notice should set out the cause of the delay, and the contractor is required to establish whether or not the cause is a 'Relevant Event'.

5.22   The contractor is also required to give particulars of the expected effects of all events identified as relevant events, and an estimate of the expected delay in the completion of the works or the section (cl 2.24.2). If it is not practical to supply this information with the notice, the contractor should identify the event as soon as possible. Following notification, the contractor must inform the employer of any change in the estimated delay or the particulars supplied (cl 2.24.3).

5.23   On receipt of the notice and particulars, the employer must assess the delay caused and make a decision regarding the appropriate extension of time, if any (cl 2.25.1). The decision must be notified to the contractor (even if the decision is that no extension is due) within 12 weeks of receipt of the notice and 'the required particulars' (cl 2.25.2). If the completion date is less than 12 weeks away, the employer shall endeavour to reach a decision prior to the completion date. Such decisions are specifically required before the provisions of Schedule 7 paragraph A.9 (the 'freezing provisions' in respect of fluctuations) can become operative.

5.24   The extension can only be given in relation to delay caused by events listed in clause 2.26. These events cover a very wide range of circumstances beyond the control of the contractor, including default by the employer and neutral events such as exceptionally adverse weather. With regard to relevant events (many of which are the same as those in SBC16) the following points should be noted:

- The contractor will be entitled to an extension if compliance with clause 3.15 (antiquities) is necessary.
- The contractor will be entitled to an extension following any exercise of its right of suspension arising from non-payment of amounts due by the employer (cl 2.26.5).
- The scope of clause 2.26.6 is very wide as it refers to 'any impediment, prevention or default, whether by act or omission, by the Employer'. This will, to an extent, duplicate some of the other events and will therefore give alternative grounds for a claim.
- Statutory undertaker's work (cl 2.26.7) covers only situations where work is carried out in pursuance of statutory duties. If the work is directly commissioned by the employer this comes under clause 2.26.6.
- Weather must be exceptional and adverse (i.e. not that which would be expected at the time of year in question) (cl 2.26.8). The effect of the weather is assessed at the time the work is actually carried out, not when scheduled in the contractor's programme.
- Specified perils (cl 2.26.9) can, under certain circumstances, include events caused by the contractor's own negligence.
- Clause 2.26.11 also includes strikes, etc., affecting those engaged in preparation of the design of the works. (Note that the general protection afforded with respect to strikes is very wide, i.e. not simply those directly affecting the works, but also those causing difficulties in preparation and transportation of goods and materials. Such strikes will not necessarily be confined to the UK and, given the current extent of overseas imports, the effects could be considerable.)
- The exercise of a statutory power by the UK Government or a local or public authority is a relevant event, provided this was not necessitated by a breach of the contractor (cl 2.26.12).
- Delay in receipt of planning or other necessary permissions or approvals of statutory bodies is a relevant event, provided that the contractor has taken all practicable steps to avoid the delay (cl 2.26.13).

- There is no ground for claiming for delay caused by a named sub-contractor.
- There is no ground for claiming due to shortage of or difficulty in obtaining labour, goods or materials.

5.25 Failure to grant an extension properly due could result in the contractor no longer being obliged to finish by the completion date (although this is unlikely to be the case if the contractor has failed to comply with the procedural requirements relating to applications, *Multiplex* v *Honeywell*). The employer should nevertheless take care to deal with the matter within the stipulated time limits. Clause 2.25.3 requires the employer to identify in any extension the relevant events which have been taken into account, and whether any account has been taken of work which has been omitted. The employer must also notify the contractor if it is decided that no extension of time is due.

> *Multiplex Constructions (UK) Limited* v *Honeywell Control Systems Limited (No. 2)* [2007] EWHC 447
>
> Wembley National Stadium Limited (WNSL) contracted with Multiplex Constructions (UK) Ltd (Multiplex), to construct the new Wembley National Stadium. Multiplex engaged Honeywell Control Systems Limited (Honeywell) under a sub-contract to design, supply and install various electronic systems. By the time Honeywell entered into the sub-contract, substantial delays to the project had already occurred. Multiplex issued three revised programmes to Honeywell, extending the completion date to 31 March 2006. The date passed without completion being achieved. No further programmes were issued by Multiplex. Honeywell maintained that the issue of the three programmes entitled it to claim prolongation costs and other financial relief. Honeywell argued that time had been set at large, due to its non-compliance with the conditions precedent. The judge decided that the contractor still had to comply with any notice provisions before an extension of time application could be entertained, even in relation to acts of prevention by the employer: 'Contractual terms requiring a contractor to give prompt notice of delay serve a valuable purpose; such notice enables matters to be investigated while they are still current'.

5.26 Following the first extension of time, under clause 2.25.4 the employer may fix an earlier completion date than that previously fixed, where there is a change which results in the omission of any work or obligation or restriction, and the employer may do so without having received any notice from the contractor. However, clause 2.25.6.3 makes it clear that the employer may not fix a date earlier than the original date for completion.

5.27 The employer may award further extensions of time in respect of relevant events which occur after the completion date but before practical completion, i.e. when the contractor is in 'culpable delay' (*Balfour Beatty* v *Chestermont Properties*). In this event, the extension is added onto the date that has passed, referred to as the 'net' method of extension. The employer may also bring forward the completion date, where appropriate, due to work being omitted.

> *Balfour Beatty Building Ltd* v *Chestermont Properties Ltd* (1993) 62 BLR 1
>
> In a contract on JCT80, the works were not completed by the revised completion date and the architect issued a non-completion certificate. The architect then issued a series of variation instructions and a further extension of time, which had the effect of fixing a completion date two-and-a-half months before the first of the variation instructions. He then issued a further

> non-completion certificate and the employer proceeded to deduct liquidated damages. The contractor took the matter to arbitration, and then appealed certain decisions on preliminary questions given by the arbitrator. The court held that the architect's power to grant an extension of time pursuant to clause 25.3.1.1 could only operate in respect of relevant events that occurred before the original or the previously fixed completion date, but the power to grant an extension under clause 25.3.3 applied to any relevant event. The architect was right to add the extension of time retrospectively (termed the 'net' method).

5.28 Following practical completion of the works, or of a section, the employer may grant a further extension of time at any point up to 12 weeks after practical completion (cl 2.25.5) and may review any extension of time previously given. The final review may extend the date previously fixed and may bring it forward where work is omitted, but not to a date earlier than the date for completion entered in the contract particulars. At this point there is no requirement for notification by the contractor, although it is likely that the review will be the subject of some correspondence between the parties. The employer must notify the contractor of the result of the review by the end of the 12-week period, including if the decision is that no further extension is due.

Assessment

5.29 The contractor is entitled to extensions of time properly due under the contract and any failure on the part of the employer to comply with the provisions would constitute a breach of contract. However, the employer has no power to grant extensions of time except as provided for in the contract, and for delays caused by events listed in clause 2.26.

5.30 The employer should make an objective assessment of every notice received. The aim is to establish whether a delay has been caused by the event cited, whether the delay is likely to disrupt the programme and, consequently, delay the final completion date and, if so, to assess the probable extent of that final delay. Any contractor's programme can be used as a guide and may be particularly useful where the programme shows a critical path, but although it may be persuasive evidence it is not conclusive or binding. The effect on progress is assessed in relation to the work being carried out at the time of the delaying event, rather than the work that was programmed to be carried out.

5.31 Clause 2.25.6.1 contains the important proviso that the contractor must 'constantly use his best endeavours to prevent delay'; therefore, the employer can assume that the contractor will take steps to minimise the effect of the delay on the completion date, for example by reprogramming the remaining work. The phrase 'best endeavours' requires something more than 'reasonable' or 'practicable', but it is unlikely to extend to excessive expenditure. Clause 2.25.6.2 states that the contractor 'shall do all that may reasonably be required to the satisfaction of the Employer to proceed with the Works'. However, if the employer requires measures to be taken which amount to a change in the employer's requirements, then this may result in a claim for loss and/or expense.

5.32 The effect on completion of any delay, taking into account the contractor's 'best endeavours' (cl 2.25.6.1), is not always easy to predict. The employer is required to form a view on the delay, and in doing this the employer would be wise to take an objective view, even where the delay has been caused by the employer. If the contractor believes that the employer has not awarded a sufficient extension, it could raise this dispute in adjudication and if the contractor is proved correct the employer will not only be obliged to grant the extension, but will also be liable for losses to the contractor.

5.33   Two or more delaying events may happen simultaneously, or with some overlap, and this can raise difficult questions with respect to the award of an extension of time. In the case of concurrent delays involving two or more relevant events, it has been customary to grant the extension in respect of the dominant reason, but this is only appropriate where the dominant reason begins before and ends after any other reasons. Even then, if the dominant reason is not a ground for a loss and expense claim, this may still be awarded in respect of the other delaying events (*H Fairweather & Co.* v *Wandsworth*).

> *H Fairweather & Co. Ltd* v *London Borough of Wandsworth* (1987) 39 BLR 106
>
> Fairweather entered into a contract with the London Borough of Wandsworth to erect 478 dwellings. The contract was on JCT63. Pipe Conduits Ltd was nominated sub-contractor for underground heating works. Disputes arose and an arbitrator was appointed who made an interim award. Leave to appeal was given on several questions of law arising out of the award. The arbitrator had found that where a delay occurred which could be ascribed to more than one event, the extension should be granted for the dominant reason. Strikes were the dominant reason, and the arbitrator had therefore granted an extension of 81 weeks, and made it clear that this reason did not carry any right to direct loss and/or expense. The court stated that an extension of time was not a condition precedent to an award of direct loss and/or expense, and that the contractor would be entitled to claim for direct loss and/or expense for other events which had contributed to the delay.

5.34   Where one overlapping delaying event is a relevant event and the other is not (in other words, one is the employer's risk and the other the contractor's), a difficult question arises as to the extension of time due. The instinctive reaction of many assessors might be to 'split the difference', given that both parties have contributed to the delay. However the more logical approach is that the contractor should be given an extension of time for the full length of delay caused by the relevant event, irrespective of the fact that, during the overlap, the contractor was also causing delay. Taking any other approach, for example splitting the overlap period and awarding only half of the extension to the contractor, could result in the contractor being subject to liquidated damages for a delay partly caused by the employer. The courts have normally adopted this analysis (see *Walter Lilly & Co. Ltd* v *Giles Mackay & DMW Ltd*). A leading Scottish case had stated that a proportional approach would be fairer (*City Inn Ltd* v *Shepherd Construction Ltd*), but this decision is not binding on English courts and was not approved in Walter Lilly.

> *Walter Lilly & Co. Ltd* v *Giles Mackay & DMW Ltd* [2012] EWHC 649 (TCC)
>
> This case concerned a contract to build Mr and Mrs Mackay's, and two other families', luxury new homes in South Kensington, London. The contract was entered into in 2004 on the JCT Standard Form of Building Contract 1998 Edition with a Contractor's Designed Portion Supplement. The total contract sum was £15.3 million, the date for completion was 23 January 2006, and liquidated damages were set at £6,400 per day. Practical completion was certified on 7 July 2008. The contractor (Walter Lilly) issued 234 notices of delay and requests for extensions of time, of which fewer than a quarter were answered. The contractor brought a claim for, among

> other things, an additional extension of time. The court awarded a full extension up to the date of practical completion. It took the opportunity to review approaches to dealing with concurrent delay, including that in the case of *Henry Boot Construction (UK) Ltd* v *Malmaison Hotel (Manchester) Ltd* (where the contractor is entitled to a full extension of time for delay caused by two or more events, provided one is an event which entitles it to an extension under the contract), and the alternative approach in the Scottish case of *City Inn Ltd* v *Shepherd Construction Ltd* (where the delay is apportioned between the events). The court decided that the former was the correct approach in this case. As part of its reasoning the court noted that there was nothing in the relevant clauses to suggest that the extension of time should be reduced if the contractor was partly to blame for the delay.

> *City Inn Ltd* v *Shepherd Construction Ltd* [2008] CILL 2537 Outer House Court of Session
>
> In considering a case involving a dispute over extensions of time under a JCT80 form of contract, the court considered earlier authorities and the principles underlying extension of time clauses, and set out several propositions. These included that, where there are several causes of delay and where a dominant cause can be identified, the assessor can use the dominant cause and set aside immaterial causes. However, where there are two causes of delay, only one of which is a contractor default, the assessor may apportion delay between the two events. The CILL editors describe this as going 'further than any recent authority' on concurrency. Assessors should note that this is a Scottish case which has not, to date, been followed in English courts.

Partial possession

5.35 The employer may take possession of completed parts of the works ahead of practical completion by the operation of clause 2.30. 'Partial possession' requires the agreement of the contractor, which cannot be unreasonably withheld. The contractor must issue a written statement to the employer identifying precisely the extent of the 'Relevant Part' and the date of possession (the 'Relevant Date'). This statement should be prepared with great care, if necessary using a drawing to illustrate the extent, and the employer should check the statement carefully. The insurers should be notified where relevant. The statement must be issued immediately after the part is taken into possession but, in practice, it would be wise to circulate the drawings and information in advance, so that the details of what will occur are clear to all parties. The partial possession may affect other operations on site, in which case it could constitute an 'impediment'. The employer should therefore consider the possible contractual consequences in terms of delay and claims for loss and/or expense.

5.36 Clause 2.31 states that practical completion is 'deemed to have occurred' for the 'Relevant Part' of the works, and that the rectification period for that part is deemed to commence on the 'Relevant Date'. The employer is required to issue a separate notice when defects have been made good (cl 2.32). However, the relevant part remains part of the works, and is still to be included under the statement of practical completion. It is notable that the clause does not state that the works to the 'Relevant Part' must have reached practical completion but, in view of the contractual consequences, it would be unwise for the employer to take possession before practical completion has been achieved (see paragraph 5.45 below).

5.37   Liquidated damages are reduced by the proportion of the value of the possessed part of the works to the contract sum (cl 2.34). The effect of clause 4.18.2 is that half of the retention is released for that proportion of the works. If Insurance Option A, B or C.2 applies, the obligation to insure ceases (cl 2.33), and the employer may wish to consider insuring that part, as the contractor's obligation to insure the works will cease.

5.38   It is important to note that the fact that significant work remains outstanding has not prevented the courts from finding that 'partial possession' has been taken of the whole works, in situations where a tenant has effectively occupied the whole building, allowing access to the contractor for remedial work (see *Skanska Construction (Regions) Ltd* v *Anglo-Amsterdam Corporation Ltd*). If the parties do not intend clause 2.25 to take effect for the whole project, they must make clear, under a carefully worded agreement, what are the contractual consequences of any intended occupation (see below).

> *Skanska Construction (Regions) Ltd* v *Anglo-Amsterdam Corporation Ltd* (2002) 84 Con LR 100
>
> Anglo-Amsterdam Corporation Ltd (AA) engaged Skanska Construction (Skanska) to construct a purpose-built office facility under a JCT81 With Contractor's Design form of contract. Clause 16 had been amended to state that practical completion would not be certified unless the certifier was satisfied that any unfinished works were 'very minimal and of a minor nature and not fundamental to the beneficial occupation of the building'. Clause 17 of the form stated that practical completion would be deemed to have occurred on the date that the employer took possession of 'any part or parts of the works'.
>
> AA wrote to Skanska confirming that the proposed tenant for the building would commence fitting-out works on the completion date. However, the air-conditioning system was not functioning and Skanska had failed to produce operating and maintenance manuals. Following this date the tenant took over responsibility for security and insurance, and Skanska was allowed access to complete outstanding work. AA alleged that Skanska was late in the completion of the works and made a claim for liquidated damages at the rate of £20,000 per week for a period of approximately nine weeks. Skanska argued that the building had achieved practical completion on time or that, alternatively, partial possession of the works had taken place and that, consequently, its liability to pay liquidated damages had ceased under clause 17.
>
> The case went to arbitration and Skanska appealed. The court was unhappy with the decision and found that clause 17.1 could also operate when possession had been taken of all parts of the works and was not limited to possession of only part or some parts of the works. Accordingly, it found that partial possession of the entirety of the works had, in fact, been taken some two months earlier than the date of practical completion, when AA agreed to the tenant commencing fit-out works. Consequently, even though significant works remained outstanding, Skanska was entitled to repayment of the liquidated damages that had already been deducted by AA.

Use or occupation before practical completion

5.39   Clause 2.5 provides for the situation where the employer wishes to 'use or occupy the site or the Works or part of them' while the contractor is still in possession. The purposes for which the employer might require such use are described simply as 'storage or otherwise' and, while this might suggest that the intention is of a limited role, in theory at least the clause places no limits on what form such use or occupation might take. In practice, as the written consent of the contractor is required and (as with partial possession discussed above) it appears neither to be included as an event for which an extension of time may be awarded, nor as a matter for which losses for disturbance can be claimed, it would not be unreasonable for the contractor to withhold permission unless it could be established that the use or occupation would cause no delays. Before the contractor is required to give consent, the employer or the contractor (as appropriate) must notify the insurers. Any additional premium required is added to the contract sum (cl 2.5.2).

5.40 Situations can arise where the contractor has failed to meet the completion date and, although no sections of the works are sufficiently complete to allow the employer to take possession of those parts under clause 2.27, the employer is, nevertheless, anxious to occupy at least part of the works. Nothing in the contract specifically allows for this situation. A suggestion was put forward in the 'Practice' section of the *RIBA Journal* (February 1992), which has frequently proved useful in practice (see Figure 5.2): in return for being allowed to occupy the premises, the employer agrees not to claim liquidated damages during the period of occupation. A statement of practical completion obviously cannot be issued, and retention money cannot be released until it is. The insurance of the works will need to be settled with the insurers.

5.41 Such an arrangement would be outside the terms of the contract. It should, therefore, be covered by a properly drafted agreement which is signed by both parties. (The cases of

---

Figure 5.2 'Practice' section, *RIBA Journal* (February 1992)

Employer's possession before practical completion under JCT contracts

It is not uncommon for the employer, after the completion date has passed, to wish to take possession of the Works before the contractor has achieved practical completion. In this event an ad hoc agreement between employer and contractor is required to deal with the situation. In respect of such an agreement, members may wish to have regard to the following note...

Outstanding items

Where it is known to the architect that there are outstanding items, practical completion should not be certified without specially agreed arrangements between the employer and the contractor. For example, in the case of a contract where the contract completion date has passed it could be so agreed that the incomplete building will be taken over for occupation, subject to postponing the release of retention and the beginning of the defects liability period until the outstanding items referred to in a list to be prepared by the architect are completed, but relieving the contractor from liability for liquidated damages for delay as from the date of occupation, and making any necessary changes in the insurance arrangements. In such circumstances either the Certificate of Practical Completion form should not be used or it should be altered to state or refer to the specially agreed arrangements. In making such arrangements the architect should have the authority of the client-employer.

When the employer is pressing for premature practical completion there is a need to be particularly careful where there are others who are entitled to rely on the issue of a Practical Completion Certificate and its consequences. In the case where part only of the Works is ready for handover the partial possession provisions can be operated to enable the employer with the consent of the contractor to take possession of the completed part.

Skanska v *Anglo-Amsterdam Corporation* above and *Impresa Castelli* v *Cola* illustrate the importance of drafting a clear agreement.) It may also be advisable to agree that, in the event of the contractor's failure to achieve practical completion by the end of a further agreed period, liquidated damages would begin to run again, possibly at a reduced rate. In most circumstances, this arrangement would be of benefit to both parties, and is certainly preferable to issuing a heavily qualified statement of practical completion listing 'except for' items.

> *Impresa Castelli SpA* v *Cola Holdings Ltd* (2002) CLJ 45
>
> Impresa agreed to build a large four-star hotel for Cola Holdings Ltd (Cola), using the JCT Standard Form of Building Contract With Contractor's Design, 1981 edition. The contract provided that the works would be completed within 19 months from the date of possession. As the work progressed, it became clear that the completion date of February 1999 was not going to be met, and the parties agreed a new date for completion in May 1999 (with the bedrooms being made available to Cola in March) and a new liquidated damages provision of £10,000 per day, as opposed to the original rate of £5,000. Once the agreement was in place, further difficulties with progress were encountered, which meant that the May 1999 completion date was also unachievable. The parties entered into a second variation agreement, which recorded that Cola would be allowed access to parts of the hotel to enable it to be fully operational by September 1999, despite certain works being incomplete (including the air conditioning). In September 1999, parts of the hotel were handed over, but Cola claimed that such parts were not properly completed. A third variation agreement was put in place with a new date for practical completion and for the imposition of liquidated damages.
>
> Disputes arose and, among other matters, Cola claimed for an entitlement for liquidated damages. Impresa argued that it had achieved partial possession of the greater part of the works, therefore a reduced rate of liquidated damages per day was due. The court found that, although each variation agreement could have used the words 'partial possession', they had in fact instead used the word 'access'. The court had to consider whether partial possession had occurred under clause 17.1 of the contract, which provides for deemed practical completion when partial possession is taken, or whether Cola's presence was merely 'use or occupation' under clause 23.3.2 of the contract. The court could find nothing in the variation agreements to suggest that partial possession had occurred. It therefore ruled that what had occurred related to use and occupation, as referred to in clause 23.3.2 of the contract, and that the agreed liquidated damages provision was therefore enforceable.

Practical completion

5.42 The employer is obliged to issue the contractor with a written statement (cl 2.27) when:

- the works (or a section) have reached practical completion (see below); and
- the contractor has 'complied sufficiently with clauses 2.37 and 3.16 in respect of the supply of documents and information' (i.e. as-built drawings and information required for the health and safety file).

5.43 Clause 2.27 then states that 'practical completion of the Works or the Section shall be deemed for all the purposes of this Contract to have taken place on the date stated in that statement'. Although the wording of clause 2.27 is somewhat circular, a reasonable interpretation of this clause is that practical completion only occurs when both conditions are met, the principal argument for this reasoning being that only one date is named in the statement. The employer would be entitled to withhold the statement until all significant as-built and health and safety information has been received, even if the actual works have been finished for some time, and would certainly be entitled to do so if the lack of information put the employer at risk of being in breach of the CDM Regulations. The use of the term 'complied sufficiently' (cl 2.27) may allow the employer to use its discretion in issuing the statement despite some information being missing.

5.44   Deciding when the works have reached practical completion often causes some difficulty. It is suggested that practical completion means the completion of all works required under the contract and by subsequent instruction. Although it has been held that an architect has discretion under JCT traditional forms to certify practical completion where there are very minor items of work left incomplete, on *de minimis* principles (*City of Westminster* v *Jarvis* and *H W Nevill (Sunblest)* v *William Press*; *Hall* v *Van Der Heiden*), the employer would be wise to proceed with caution. Contrary to the opinion of many contractors, there is no obligation to issue the statement when the project is 'substantially' complete (or even when it is capable of occupation by the employer) if items are still outstanding.

> *City of Westminster* v *J Jarvis & Sons Ltd* (1970) 7 BLR 64 (HL)
>
> Jarvis entered into a contract with the City of Westminster to construct a multi-storey car park in Rochester Row. The contract was on JCT63, which included 'delay on the part of a nominated sub-contractor ... which the Contractor has taken all reasonable steps to avoid' (cl 23(g)). Subsequently, defects were discovered in many of the piles and the remedial works caused a delay to the contractor of over 21 weeks. The House of Lords found that, on a proper interpretation of clause 23(g), delay could be attributed to the nominated sub-contractor only if it had failed to complete its work by the date given in the sub-contract. The clause did not apply after the works had been accepted as complete. In addressing the question of whether there was a delay in completion, the court had also to consider what was meant by 'practical completion' of the sub-contract works. Viscount Dilhourne stated: 'The contract does not define what is meant by "practically completed". One would normally say that a task was practically completed when it was almost but not entirely finished; but "practical completion" suggests that that is not the intended meaning and that what is meant is the completion of all the construction work that has to be done' (at page 75).

> *H W Nevill (Sunblest) Ltd* v *William Press & Son Ltd* (1981) 20 BLR 78
>
> William Press entered into a contract with Sunblest to carry out foundations, groundworks and drainage for a new bakery on a JCT63 contract. A practical completion certificate was issued and new contractors commenced a separate contract to construct the bakery. A certificate of making good defects and a final certificate were then issued for the first contract, following which it was discovered that the drains and the hardstanding were defective. William Press returned to site and remedied the defects, but the second contract was delayed by four weeks and Sunblest suffered damages as a result. It commenced proceedings, claiming that William Press was in breach of contract and, in its defence, William Press argued that the plaintiff was precluded from bringing the claim by the conclusive effect of the final certificate. Judge Newey decided that the final certificate did not act as a bar to claims for consequential loss. In reaching this decision, he considered the meaning and effect of the certificate of practical completion and stated 'I think that the word "practically" in clause 15(1) gave the architect a discretion to certify that William Press had fulfilled its obligation under clause 21(1) where very minor de-minimis work had not been carried out, but that if there were any patent defects in what William Press had done then the architect could not have issued a certificate of practical completion' (at page 87).

5.45　An early statement of practical completion should be treated with extreme caution as it can give rise to considerable complications. Even though the employer may be keen to move into the newly completed works, and the contractor, anxious to avoid liquidated damages, may be even more enthusiastic, the temptation to issue the statement early should be avoided. The employer may later find itself in a difficult position contractually as the following areas of potential contractual difficulty come into play:

- half of the retention will be released, leaving only half in hand (cl 4.18.2). This puts the employer at considerable risk, as the 1.5 per cent remaining is only intended to cover latent defects;
- the rectification period begins (cl 2.35);
- the onus shifts to the employer to notify the contractor of all necessary outstanding work under clause 2.35. If the employer should fail to include any item the contractor would have no authority to enter the site to complete it – therefore the employer will inevitably become involved in managing and programming the outstanding work;
- possession of the site passes to the employer and, depending on the insurance arrangements, the contractor might no longer cover the insurance of the works. The insurers will need to be informed about the programme for the outstanding works;
- the contractor's liability for liquidated damages ends;
- the employer will be the 'occupier' for the purposes of the Occupiers' Liability Acts 1957 and 1984 and may also be subject to health and safety claims.

5.46　The statement must be issued as soon as the criteria in clause 2.27 are met. The contractor is obliged to complete 'on or before' the completion date and once practical completion is achieved the employer is obliged to accept the works. Employers who wish to accept the works only on the date given in the contract will need to amend the wording. If the sectional completion modifications are used, a statement of practical completion must be issued for each section of the works.

Procedure at practical completion

5.47　The contract sets out no procedural requirements for what must happen at practical completion, it simply requires the employer to issue a statement. It would be advisable for the employer's requirements to set out a procedure, particularly in relation to provisions for testing and commissioning and the satisfaction of performance requirements.

5.48　Leading up to practical completion, it is advisable for the employer (and any employer's agent) not to become too deeply involved with events on site. It sometimes happens that the employer's agent feels obliged to issue detailed 'snagging' lists of defective and outstanding work. Under the contract, responsibility for quality control and snagging rests entirely with the contractor. In adopting this role the employer may be assisting the contractor, and although this may appear to be of immediate benefit in alerting the contractor to problems, it may lead to confusion over the liability position, which might give rise to future difficulties. If the employer feels that the works are not complete, but the contractor is disputing this fact, the best course may be to draw attention to typical problems, but to make it clear that the list is indicative and not comprehensive.

5.49   It is common practice for the contractor to arrange a 'handover' meeting. The term is not used in DB16, although handover meetings can be particularly useful in design and build, where the employer will rely more heavily on the contractor to explain the operation and maintenance of the building than might be the case with traditional procurement. Nevertheless, such a meeting, regardless of its practical value, would be of no contractual significance.

Failure to complete by the completion date

5.50   In the event of failure to complete the works by the completion date, the employer is required to notify the contractor of the failure (a 'Non-Completion Notice', cl 2.28) and the notification is a condition precedent to the deduction of liquidated damages (cl 2.29.1.1).

5.51   If the sectional completion provisions are used, a separate notice will be needed for each incomplete section. Once the notice has been issued, the contractor is said to be in 'culpable delay'. The employer, provided that it has issued the necessary notices (see paragraph 5.55 below), may then deduct the damages from the next interim payment, or reclaim the sum as a debt. Note that fluctuations provisions are frozen from this point. If a new completion date is set at a later date, this has the effect of cancelling the notice and the employer must issue a further non-completion notice, if necessary.

Liquidated and ascertained damages

5.52   The agreed rate for liquidated and ascertained damages is entered into the contract particulars. This is normally expressed as a specific sum per week (or other unit) of delay, to be allowed by the contractor in the event of failure to complete by the completion date (note that several different rates may apply where the works are divided into sections). As a result of two decisions in the Supreme Court, it is no longer considered essential that the amount is calculated on the basis of a genuine pre-estimate of the loss likely to be suffered (*Cavendish Square Holdings* v *El Makdessi* and *ParkingEye Limited* v *Beavis*; see also *Alfred McAlpine Capital Projects* v *Tilebox*). Provided that the amount is not 'out of all proportion' to the likely losses, the damages will be recoverable without the need to prove the actual loss suffered, irrespective of whether the actual loss is significantly less or more than the recoverable sum (*BFI Group of Companies* v *DCB Integration Systems*). In other words, once the rate has been agreed, both parties are bound by it. Of course, for practical reasons, the rate should always be discussed with the employer before inclusion in the tender documents, and an amount that will provide adequate compensation included to cover, among other things, any additional professional fees that may be charged during this period. If 'nil' is inserted into the contract particulars then this may preclude the employer from claiming any damages at all (*Temloc* v *Errill*), whereas if no sum is entered the employer may be able to claim general damages.

> *Alfred McAlpine Capital Projects Ltd* v *Tilebox Ltd* [2005] BLR 271
>
> This case contains a useful summary of the law relating to the distinction between liquidated damages and penalties. A WCD98 contract contained a liquidated damages provision in the sum of £45,000 per week. On the facts, this was a genuine pre-estimate of loss and the actual loss suffered by the developer, Tilebox, was higher. The contractor therefore failed to obtain

a declaration that the provision was a penalty. However, the judge also considered a different (hypothetical) interpretation of the facts whereby it was most unlikely, although just conceivable, that the total weekly loss would be as high as £45,000. In this situation also the judge considered that the provision would not constitute a penalty. In reaching this decision he took into account the facts that the amount of loss was difficult to predict, that the figure was a genuine attempt to estimate losses, that the figure was discussed at the time that the contract was formed and that the parties were, at that time, represented by lawyers.

Cavendish Square Holdings v *El Makdessi* and *ParkingEye Limited* v *Beavis*, Supreme Court 2015

In this landmark case the Supreme Court restated the law regarding whether a liquidated damages clause may be considered a penalty. Key criteria for whether a provision will be penal are: if 'the sum stipulated for is extravagant and unconscionable in amount in comparison with the greatest loss that could conceivably be proved to have followed from the breach'; and whether the sum imposes a detriment on the contract breaker which is 'out of all proportion to any legitimate interest of the innocent party'. In determining these, the court must consider the wider commercial context.

BFI Group of Companies Ltd v *DCB Integration Systems Ltd* [1987] CILL 348

BFI employed DCB on the Agreement for Minor Building Works to refurbish and alter offices and workshops at its transport depot. BFI was given possession of the building on the extended date for completion, but two of the six vehicle bays could not be used for another six weeks as the roller shutters had not yet been installed. Disputes arose which were taken to arbitration. The arbitrator found that the delay in completing the two bays did not cause BFI any loss of revenue, and that BFI was therefore not entitled to any of the liquidated damages. BFI was given leave to appeal to the High Court. HH Judge John Davies QC found that BFI was entitled to liquidated damages. It was quite irrelevant to consider whether in fact there was any loss. Liquidated damages do not run until possession is given to the employer but until practical completion is achieved, which may not be at the same time. Therefore, the fact that the employer had use of the building was also not relevant.

Temloc Ltd v *Errill Properties Ltd* (1987) 39 BLR 30 (CA)

Temloc entered into a contract with Errill Properties to construct a development near Plymouth. The contract was on JCT80 and was in the value of £840,000. '£ nil' was entered into the contract particulars against clause 24.2, liquidated and ascertained damages. Practical completion was certified around six weeks later than the revised date for completion. Temloc brought a claim against Errill Properties for non-payment of certain certified amounts, and Errill counterclaimed for damages for late completion. It was held by the court that the effect of '£ nil' was not that the clause should be disregarded (because, for example, it indicated that it had not been possible to assess a rate in advance), but that it had been agreed that no damages would be payable in the event of late completion. Clause 24 is an exhaustive remedy and covers all losses normally attributable to a failure to complete on time. The defendant could not, therefore, fall back on the common law remedy of general damages for breach of contract.

5.53 Before liquidated damages may be claimed, the following preconditions must be met:

(a) the contractor must have failed to complete the works by the completion date;

(b) the employer must have fulfilled all duties with respect to the award of an extension of time;

(c) the employer must have issued a non-completion notice to the contractor (cl 2.29.1.1);

(d) the employer must have notified the contractor before the due date for the final payment under clause 4.24.5 that it may require payment of, or withhold, liquidated damages (cl 2.29.1.2).

5.54 If these preconditions are met then the employer may, not later than five days before the final date for payment of 'the amount payable under clause 4.24, issue a notice in accordance with clause 2.29.2 (cl 2.29.1). It should be noted that, although clause 2.29.1 states the employer 'may' give this notice, in effect the notice *must* be issued if the employer wishes to claim liquidated damages.

5.55 Clause 2.29.1 therefore requires two types of notice: first, a general notice of intention (cl 2.29.1.2) and then a notice not later than five days before the final payment date for the final payment (cl 2.29.1). This second notice requires more detail than the first; it must state that 'for the period between the Completion Date and the date of practical completion' the employer requires the contractor to pay the sum to the employer and/or intends to deduct liquidated damages from monies due, and whether the rate will be the contractual one or a lesser sum (cl 2.29.2). The requirement and notification must be reasonably clear, but there is no need for a great deal of detail (*Finnegan* v *Community Housing Association*). It may be possible for the clause 2.29.1.2 requirement and the clause 2.29.1 notice to be dealt with together, provided the document contains the necessary information and is issued at the right time. However, it is important to note that if the employer wishes to withhold or deduct all or any of the liquidated damages payable, footnote [37] to clause 2.29.2.2 explains that, in addition to the notice under clause 2.29.1, the employer must give the appropriate 'Pay Less Notice' under clause 4.9 (see paragraph 8.30 of this Guide). This will explain the calculation for that particular deduction, i.e. the period over which the damages are claimed, and the rate applied.

> *J F Finnegan Ltd* v *Community Housing Association Ltd* (1995) 77 BLR 22 (CA)
>
> Finnegan Ltd was employed by the Housing Association to build 18 flats at Coram Street, West London. The contractor failed to complete the work on time and the contract administrator issued a certificate of non-completion. Following the certificate of practical completion, an interim certificate was issued. The employer sent a notice with the cheque honouring the certificate, which gave minimal information (i.e. not indicating how liquidated and ascertained damages (LADs) had been calculated). The Court of Appeal considered this sufficient to satisfy the requirement for the employer's written notice in clause 24.2.1. Peter Gibson LJ stated (at page 33):
>
>> I consider that there are only two matters which must be contained in the written requirement. One is whether the employer is claiming a payment or a deduction in respect of LADs. The other is whether the requirement relates to the whole or a part (and, if so, what part) of the sum for the LADs.

> He then stated (at page 35):
>
>> I would be reluctant to import into this commercial agreement technical requirements which may be desirable but which are not required by the language of the clause and are not absolutely necessary.
>
> The requirements relating to notices have now changed. However, there appears to be no reason why the general comments would not still apply, i.e. that the amount of information required would be no more than the minimum set out in the contractual provisions.

5.56 The employer is entitled to deduct liquidated damages for the period between the completion date and the date of practical completion (cl 2.29.2). Use of the phrase 'fails to complete the Works' in clause 2.28 is unfortunate as this is only one of the criteria that has to be met before the employer issues the notice of practical completion (see paragraph 5.43), and the phrase creates certain ambiguity regarding liquidated damages. For example, a contractor might try to argue on the basis of this wording that, once it has completed the works, the employer loses the right to deduct further liquidated damages, even though the employer has not yet issued the practical completion statement. It is suggested that clause 2.28 would be much clearer if it were to read: 'If practical completion has not been achieved by the Completion Date'. Nevertheless, given the clear intention behind the clause, it is to be hoped that no such argument as that set out above would succeed.

5.57 If an extension of time is given following the issue of a notice of non-completion, then this has the effect of cancelling that notice. A new non-completion notice must be issued by the employer if the contractor then fails to complete by the revised completion date (cl 2.28). The employer must, if necessary, repay any liquidated damages recovered for the period up to the new completion date (cl 2.29.3) and must do so within a reasonable period of time (*Reinwood* v *L Brown & Sons*). Clause 2.29.4 states that any notice which has previously been issued in accordance with clause 2.29.1.2 shall remain effective 'unless the Employer states otherwise in writing', notwithstanding that a further extension of time has been granted.

> *Reinwood Ltd* v *L Brown & Sons Ltd* [2008] BLR 219 (CA)
>
> This dispute concerned a contract on JCT98, with a date for completion of 18 October 2004, and LADs at the rate of £13,000 per week. The project was delayed and, on 7 December 2005, the contractor made an application for an extension of time. On 14 December 2005, the contract architect issued a certificate of non-completion under clause 24.1. On 11 January 2006, the architect issued interim certificate no. 29 showing the net amount for payment as £187,988. The final date for payment was 25 January 2006.
>
> On 17 January 2006, the employer issued notices under clauses 24.2 and 30.1.1.3 of its intention to withhold £61,629 LADs, and the employer duly paid £126,359 on 20 January 2006.
>
> On 23 January 2006, the architect granted an extension of time until 10 January 2006, following which the contractor wrote to the employer stating that the effect of the extension of time and revision of the completion date was that the employer was now entitled to withhold no more than £12,326. The amount due under interim certificate no. 29 was, therefore, £175,662. Subsequently,

the contractor determined the contract, relying partly on the late repayment of the balance by the employer.

The appeal was conducted on the issue of whether the cancellation of the certificate of non-completion by the grant of an extension of time meant that the employer could no longer justify a deduction of LADs. The employer's appeal was allowed. The judge stated that: 'If the conditions for the deduction of LADs from a payment certificate are satisfied at the time when the Employer gives notice of intention to deduct, then the Employer is entitled to deduct the amount of LADs specified in the notice, even if the certificate of non-completion is cancelled by the subsequent grant of an extension of time.' The employer must, however, repay the additional amount deducted within a reasonable time.

5.58   In *Department of Environment for Northern Ireland* v *Farrans*, it was decided that the contractor has the right to interest on any repaid liquidated damages. This decision was criticised at the time[1] and would be unlikely to be applied in relation to DB16, where the wording is now different, but the right to interest nevertheless remains an open question.

Department of Environment for Northern Ireland v *Farrans (Construction) Ltd* (1981) 19 BLR 1 (NI)

Farrans was employed to build an office block under JCT63. The original date for completion was 24 May 1975, but this was subsequently extended to 3 November 1977. During the course of the contract the architect issued four certificates of non-completion. By 18 July 1977, the employer had deducted £197,000 in liquidated damages but, following the second non-completion certificate, repaid £77,900 of those deductions. This process was repeated following the issue of the subsequent non-completion certificates. Farrans brought proceedings in the High Court of Justice in Northern Ireland, claiming interest on the sums that had subsequently been repaid. The court found for the contractor, stating that the employer had been in breach of contract in deducting monies on the basis of the first, second and third certificates, and that the contractor was entitled to interest as a result. The BLR commentary should be noted, which questions whether a deduction of liquidated damages permitted by clause 24.2 can be considered a breach of contract retrospectively. However, the case has not been overruled.

---

[1] See the commentary in volume 19 of the Building Law Reports.

6 Control of the works

6.1 In DB16 there is no reference to an architect or contract administrator, and both the execution of the works and many aspects of the administration of the contract lie in the hands of the contractor. This reflects the intention in design and build procurement that the employer will have less involvement with day-to-day aspects of the contract than might be the case in traditional procurement, and that the contractor will shoulder a greater responsibility for overall co-ordination of the project. The employer, for example, is not required to produce any further design information after the contract is entered into. The employer is nevertheless required at various stages in the contract to issue instructions, notifications, consents and decisions, and is likely to appoint consultants to give advice on these.

Employer's agent

6.2 Under Article 3, the employer is entitled to appoint a person to act as employer's agent, and to remove and replace the agent at will, provided it notifies the contractor of the agent's identity. The agent is not referred to again in the conditions, except for clause 2.7.3 (availability of documents for inspection) and clause 3.1 (access to the site). If an agent is named, however, the contractor would be obliged and entitled to treat the agent as the employer for all purposes of the contract; for example, applications for payment would be submitted to the agent. This would be the case in relation to all references to the 'Employer' in the conditions, unless the employer informs the contractor otherwise by written notice under Article 3. Although clauses 2.7.3 and 3.1 make specific reference to the employer's agent, and not the employer, it is likely that the clauses would be interpreted to include the employer. However, if the employer does not wish to employ an agent, then it should enter 'the Employer' in Article 3, rather than leaving the article blank or deleting it as, otherwise, it is not entirely clear what the effect of these clauses would be.

Site manager and contractor's persons

6.3 The contractor is required to keep a competent 'Site Manager' on the site 'at all material times' (cl 3.2). There is no requirement in the contract conditions for the person to be named, nor for the contractor to seek approval of the employer before appointing or replacing them. Nevertheless, this is an important role, as this person may receive instructions from the employer, and therefore acts as the contractor's agent. It would therefore be good practice to establish the identity of the site manager in a pre-contract meeting, and to make sure this is recorded in writing. The contractor has full responsibility for the performance of the site manager and all people the site manager engages on the project, and the employer is given no power to remove or replace any of these people (in practice this right is often added through amendments to the contract).

6.4 Under clause 3.1 the contractor is required to allow access to the site and to workshops to the employer's agent and any person authorised by the employer (which could also include a clerk of works). This requirement is very broadly expressed, and would even appear to cover the permanent on-site presence of a person authorised by the employer. It does not require the details of any authorised person's access to be set out in the employer's requirements, although it may be sensible to include this information if it is available. The only restrictions that may be imposed are 'as necessary to protect proprietary rights' of the contractor or any sub-contractor.

Principal contractor

6.5 The contract assumes that the contractor will act as principal contractor for the purposes of the CDM Regulations, unless another firm is named in Article 6. It is the employer's responsibility to appoint a principal contractor, therefore if the contractor is unable to or ceases to take on this role, the employer must appoint a substitute (Article 6). It is the contractor's responsibility to develop the construction phase plan so that it complies with the Regulations, and to ensure that the works are carried out in accordance with the plan.

Principal designer

6.6 The principal designer (if not the contractor, see paragraph 4.28) has no duty to inspect the works and would be very unlikely to visit the site unless some very unusual circumstance arises, such as the discovery of an unanticipated hazard. The main responsibility for ensuring that correct health and safety measures are employed on site rests with the contractor, both under statute and under the express terms of the contract (cl 3.16.2).

Employer's obligations

6.7 DB16 refers in only one place to the employer's obligation to provide information, that obligation being to define the site boundaries (cl 2.9). However, it should be noted that failure to provide any necessary instructions, decisions, information or consents may constitute a default by the employer for which an extension of time may be granted (cl 2.26.6) and may give rise to a direct loss and/or expense claim where the failure causes disruption to progress (cl 4.21.2). It is also grounds for termination if the failure causes a suspension in the works for a period greater than the amount stated in the contract particulars. In all these cases the failure constitutes grounds only when it was not caused by the negligence or default of the contractor; an example of such negligence might be failing to notify the employer in reasonable time of the need to take a decision.

6.8 There are many places in the contract where the employer is required (or has the power) to make a decision, issue an instruction, consent to an action or notify the contractor in writing regarding a matter. The obligation or power is expressed in a variety of different ways in different clauses, but the effect of failure to comply is the same, and the employer will need to be prepared to deal with any such issues promptly. Clause 1.10 now requires that all 'consents or approvals' shall not be unreasonably delayed or withheld (except in relation to clause 7.1, assignment). The key instances of such decisions, consents, etc. are highlighted in Tables 6.1 and 6.2.

Table 6.1 Key obligations of the employer

Clause	
2.3	Give possession of the site to contractor
2.5.1	If responsible for works insurance, notify the insurers regarding use or occupation of site
2.7.2	Provide contractor with certified copy of contract documents, together with pre-construction information required under the CDM Regulations
2.7.4	Not divulge information
2.9	Define the boundaries of the site
2.10.1	Instruct correction of errors in the definition of site boundary
2.10.2	Give written notice specifying divergence, if found
2.13	Issue instructions with regard to discrepancies, if found
2.14.1	Respond in writing to the contractor's proposals to deal with discrepancies
2.14.2	Decide between discrepant items or accept contractor's proposal
2.15.1	Give notice of divergences between statutory requirements and employer's requirements or any change, and consent to contractor's proposed amendment
2.15.2.3	Issue instructions requiring a change
2.20.2	Notify the contractor if use of documents may infringe patent rights
2.25.1	Give extensions of time
2.25.2	Notify the contractor of the decision regarding an extension of time
2.25.3	State the extension of time for each relevant event and reduction for each relevant omission
2.25.5	Review extensions of time previously given
2.27	Issue a practical completion statement
2.28	Issue a non-completion notice
2.29.3	Pay or repay liquidated damages to the contractor
2.32	Issue a notice that defects have been made good
2.35.1	Prepare and deliver schedule of defects
2.36	Issue 'Notice of Completion of Making Good'
3.8	Respond to contractor queries regarding basis of instructions
3.9.4	Vary the terms of the instruction to remove the contractor's objection
3.11	Instruct expenditure of provisional sums
3.15.2	Issue instructions regarding antiquities on site
3.16.1	Ensure the principal designer and the principal contractor carry out their CDM duties, where the contractor does not undertake these roles
3.16.5	Notify the contractor of any appointment of a replacement principal designer or principal contractor
4.6	Make advance payments to contractor
4.7.1	Make interim payments to contractor

Table 6.1 Key obligations of the employer – Continued

Clause	
4.7.5	Give a payment notice to the contractor specifying the amount due
4.8	Give a final payment notice to the contractor specifying the amount due
4.9.2	Pay the contractor the amount stated as due in the payment notice
4.9.3	Pay the contractor the amount stated as due in the interim payment application
4.9.4	Pay the contractor the amount stated as due in the final payment notice
4.9.5	Give a pay less notice to the contractor specifying amount to be withheld
4.9.6	Pay interest to the contractor
4.16.2	Place interest in a separate designated banking account
4.20.4	Notify the contractor of the ascertained amount of any direct loss and/or expense
6.7.2	Maintain a joint names insurance policy for the works, if responsible for this under the applicable option
6.9	Ensure insurance policy covers sub-contractors
6.10.1	Take out terrorism cover as extension to insurance policy (Insurance Option B or C)
6.11.1	Give notice if availability of terrorism cover ceases
6.12.1	Provide evidence of insurance to contractor
6.13.5.1	Pay insurance monies received to the contractor
6.18–6.19	Comply with the Joint Fire Code
8.7.4	Prepare an account
8.7.5	Pay the contractor any amount due after completion of the works by others
8.8.1	Notify contractor of decision not to complete works, prepare an account and pay the contractor any amount due
8.10.2	Inform the contractor of any insolvency event
8.12.3	Prepare an account and pay the contractor the amount due
8.12.5	Pay the contractor the amount properly due
9.1	Give serious consideration to any request for mediation
Schedule 1:	
2	Comment on and return contractor's design documents
Schedule 2:	
1.1.2	Remove reason, direct the contractor to carry out the work or omit named sub-contractor work
2.4	Take all reasonable steps to agree estimates
3.4	Give notice of decision regarding contractor's estimate of direct loss and/or expense
4.3	Confirm acceptance or otherwise of an acceleration quotation
5	Work in a collaborative manner with the contractor and other team members

Table 6.1 Key obligations of the employer – Continued

Clause	
6.1	Endeavour to maintain a working environment in which health and safety is of paramount concern
7.3	Confirm any agreed cost-saving changes with an instruction
9.1	Monitor and assess the contractor's performance against performance indicators
10	Notify the contractor promptly of any matter likely to give rise to a dispute and meet to negotiate in good faith to resolve the matter
11	Determine whether any of the content of the contract is exempt from disclosure, inform the contractor of any request for disclosure
Schedule 3:	
B.1	Take out 'All Risks' insurance of work
C.1	Insure structure against specified perils
C.2	Insure works against all risks
Schedule 4:	Endeavour to agree the amount of opening up and testing

Table 6.2 Key powers of the employer

Clause	
2.2.1	Consent to proposed substitution of materials, goods or items by contractor
2.2.4	Require contractor to produce reasonable proof that materials and goods comply with the contract requirements
2.4	Defer possession
2.5.1	Use or occupy the site
2.6	Engage persons directly to carry out work
2.7.4	Use documents supplied by the contractor for maintenance, advertisement, etc.
2.15.1	Consent to contractor's proposal for dealing with divergences
2.21	Consent to the removal of unfixed materials and goods
2.25.4	Fix earlier completion date following omission instruction
2.25.5	Review completion date
2.29.1	Require payment of liquidated damages
2.30	Take partial possession of the works
2.35	Issue instructions that defects are not to be made good
2.35.2	Instruct that any defects are to be made good
3.3.1	Consent to sub-letting of the works
3.3.2	Consent to sub-letting of the design

Table 6.2 Key powers of the employer – Continued

Clause	
3.6	Employ others where contractor fails to comply with instructions
3.7.1, 3.7.3	Confirm oral instructions in writing
3.9.1	Issue instructions effecting a change in the employer's requirements
3.10	Issue instructions requiring the contractor to postpone any of the design or construction work
3.12	Issue instructions requiring opening up or tests
3.13	In respect of work, materials or goods which are not in accordance with the contract: instruct removal from site, issue change instructions, issue instructions for further tests
3.14	Issue instructions requiring a change to deal with non-compliant workmanship
4.16	Deduct retention
4.24.4	Issue a final statement
6.12.2	Insure against risks exposed to as a result of contractor's failure to take out insurance, deduct amounts payable from sums due
6.13.5.2	Retain amounts to cover professional fees from insurance monies
6.14	Terminate the contractor's employment
6.19.2	Employ other persons to carry out remedial measures
7.1	Assign the contract with the contractor's consent
7.2	Assign limited rights under the contract
7C	Require the contractor to enter into a collateral warranty with a purchaser or tenant
7D	Require the contractor to enter into a collateral warranty with a funder
8.4.1	Give notice specifying defaults
8.4.2, 8.5.1, 8.6, 8.11	Terminate employment of contractor
8.5.3.3	Take reasonable measures to ensure the site and works are protected
8.7.1	Employ others to complete the works
8.7.2.3	Pay sub-contractors and suppliers directly
8.8.1	Decide not to have the works completed
Schedule 1:	
1.2	Comment on drawings, etc. submitted by contractor
Schedule 2:	
2.5	Instruct that Supplemental Provision 2 shall not apply or withdraw instruction
3.4	Accept estimate, negotiate estimate, instruct that clause 4.20 shall apply
4.1.4	Seek revised proposals relating to an acceleration quotation
11.3	Inform the contractor that performance indicator targets may not be met

Information to be provided by the contractor

6.9 The contractor as 'Principal Contractor' may be required by the principal designer to provide information in relation to the health and safety file. If the contractor is acting as principal designer then, under clause 3.16.2, it is required to prepare and deliver the health and safety file to the employer.

6.10 Under clause 2.37, the contractor is obliged to provide the employer before practical completion with 'such Contractor's Design Documents and related information as specified in the Contract Documents and as the Employer may reasonably require'. These documents and information are those which 'show or describe the Works as built or relate to the maintenance and operation of them'. Note that if nothing is set out in the contract documents the contractor is still under an obligation to provide such information as may reasonably be required, which in the case of most buildings could amount to a substantial operation and maintenance manual. Given that this is a design and build contract, it would be expected that the contractor will provide full details of maintenance of mechanical services, the building envelope, etc. Nevertheless, to avoid arguments about what might be reasonable, it would be sensible to give details of what will be required in the contract. The employer would be able to withhold the statement of practical completion until this information has been provided (cl 2.27).

6.11 Under clause 2.2.3 the contractor must provide such samples of the standard or quality of materials, goods and workmanship as are set out in either the employer's requirements or the contractor's proposals. The contract does not state what should happen if the employer is not satisfied with the samples provided. If the samples are not in accordance with any specification included in the requirements, then this could be pointed out to the contractor, with the disclaimer that the rejection (or acceptance) of the sample does not relieve the contractor of its obligations. If the sample does not in fact contravene anything in the requirements, then the only course will be to instruct a change in the requirements in order to stipulate an acceptable standard. This instruction may entitle the contractor to claim an extension of time and loss and/or expense, as might any delay by the employer in making a decision or giving any instruction. It should be noted that there is no right to comment on the samples, or any procedure or time interval specified during which the employer may make comments or reach any decision regarding samples. It would be possible to amend the clause to refer to specific provisions set out in the requirements (along the lines of Schedule 1). However, provided the samples meet the standards set out in the requirements, there should be no need to comment – any attempts to raise or alter the standard is likely to constitute a change.

Design submission procedure

6.12 DB16 contains detailed provisions regarding the submission of the developing design by the contractor. This information is essential in order for the employer to monitor the development of the design and to integrate it with the rest of the works.

6.13 The contractor must provide to the employer copies of the 'Contractor's Design Documents' (cl 2.8). 'Contractor's Design Documents' are defined as 'the drawings, details and specifications of materials, goods and workmanship and other related documents and information prepared by or for the Contractor in relation to the design of the Works (including such as are contained in the Contractor's Proposals), together, where applicable, with any other design documents or information to be provided by him under the BIM Protocol' (cl 1.1). It should be noted that the definition refers to 'information prepared by the Contractor' and clause 2.8 (unlike clause 2.37) does not refer to 'information reasonably required by the Employer'. The information the contractor may prepare to construct the work may be different from the information that the employer would like to receive. If specific information is needed, and the BIM protocol is not used, then it may be sensible to set out a schedule in the employer's requirements.

6.14   As regards timing, the information is to be provided 'as and when necessary from time to time in accordance with the Design Submission Procedure', and 'the Contractor shall not commence any work to which such a document relates before that procedure has been complied with' (cl 2.8). The 'Design Submission Procedure' is defined in clause 1.1 as 'such procedure as is specified in the BIM Protocol or, where that is not applicable, the procedure set out in Schedule 1, subject to any modifications of that procedure set out in the Contract Documents'.

Schedule 1 procedure

6.15   The design submission procedure (Schedule 1) states that the documents should be submitted 'by the means and in the format stated in the Employer's Requirements' and also states 'and in sufficient time to allow any comments of the Employer to be incorporated' (Schedule 1:1). It would therefore be open to the employer, and on most projects would be wise, to include detailed requirements regarding format and timing of submissions in the contract documents.

6.16   Following submission of a design document, the employer must respond within 14 days of the date of receipt, 'or (if later) 14 days from either the date or expiry of the period for submission of the same stated in the Contract Documents' (Schedule 1:2). In other words, if the contractor supplies information earlier than any agreed date, the employer is not required to respond any earlier than the date stated in the contract documents.

6.17   The employer is entitled to take three alternative courses of action: it can accept the design document, in which case it should return it marked 'A'; it may accept it, subject to certain comments being incorporated, in which case it should be marked 'B'; or it can make comments and require the contractor to resubmit the document with the comments incorporated for further approval, in which case it should be marked 'C' (Schedule 1:5). In the cases of 'B' or 'C', the employer must state why the document does not comply with the contract. (Comments are only valid if the document does not comply (Schedule 1:4); if it does comply, any required alteration would constitute a change.) If the employer does not respond within the specified period, it is deemed to have accepted the document (Schedule 1:3).

6.18   Schedule 1 paragraph 8.3 states that neither any comments nor any action by the employer will relieve the contractor of its liability to ensure that the document complies with the contract, or that the project complies with the contract. This has the effect that, if the contractor incorporates a comment made by the employer, then it accepts that the comment has been properly made (i.e. it identifies a way in which the design document is not in accordance with the contract).

6.19   If the contractor disagrees with a comment and considers that the document complies with the contract, it is required to inform the employer, within seven days of receipt of the comment, that compliance with the comment would give rise to a change (Schedule 1:7). The contractor must explain the reasoning behind its conclusions. The employer must either confirm or withdraw the comment within seven days. The confirmation or withdrawal does not signify that the employer accepts that the design document complies, or that the comment represents a change (Schedule 1:8.1); this would be a question of fact, to be resolved by adjudication if necessary. The contractor would have to implement the comment and argue its case later.

6.20 If the contractor does not notify the employer of its disagreement with a comment, then that comment will not be treated as a change, even if it could, in fact, later be shown to be a change (Schedule 1.8.2).

Employer's instructions

6.21 The employer has the power to issue instructions regarding the matters shown in Table 6.3. Only the employer (or the employer's agent) has the power to issue instructions.

Table 6.3 Matters over which the employer may or shall issue instructions

Clause	
2.10.1	Correcting errors in definition of site boundary
2.13	Correcting discrepancies between documents
2.15.2.3	Requiring a change
2.35	To not make good a defect
2.35.2	To make good defects during the rectification period
3.9.1	Effecting a change in the employer's requirements
3.10	Postponement of work
3.11	Requiring expenditure of provisional sums
3.12	Requiring opening up or tests
3.13.1	Requiring removal from site of work, materials or goods which are not in accordance with the contract
3.13.2	Requiring a change reasonably necessary as a consequence of removal
3.13.3	To open up work or carry out further tests
3.14	Requiring a change or otherwise, where contractor fails to carry out work in a proper and workmanlike manner
3.15.2	Concerning antiquities on site
Schedule 2:	
1.1.2	Remove reason or direct the contractor to carry out the work or omit named sub-contractor work
2.5.1	That the contractor complies with an instruction
4.3	Acceptance of an acceleration quotation, including change to contract sum and completion date
7.3	Agreement of cost-saving changes, including change to contract sum and completion date

Sometimes the contract states that the employer 'may' issue instructions (e.g. instructions requiring a change under clause 3.9) but, at other times, the employer 'shall' issue instructions (e.g. instructions regarding provisional sums under clause 3.11), which has the force of an obligation.

6.22   If the employer has appointed an agent, then the agent would have authority to give an instruction. If anyone other than the employer or an agent were to give an instruction, this would not be effective under the contract and the contractor would be under no obligation (and would be unwise) to comply with any such instruction.

6.23   Clause 1.7.1 states that all instructions must be in writing. An oral instruction from the employer is of no immediate effect (cl 3.7.1). However, the contractor is obliged to confirm its terms in writing within seven days. The instruction takes effect seven days from the receipt of the confirmation by the employer, provided that the employer does not dissent by notice within seven days (cl 3.7.1). If the employer confirms the instruction prior to the seven-day period elapsing, the instruction takes effect from the date of the employer's confirmation (cl 3.7.2). If neither party confirms the oral instruction, but the contractor acts upon it anyway, the employer is given the option of later confirming the instruction at any point up until the due date for the final payment under clause 4.24.5 (cl 3.7.3). However, the contractor would have taken a risk (*MOD* v *Scott Wilson Kirkpatrick*).

Ministry of Defence v *Scott Wilson Kirkpatrick & Partners* [2000] BLR 20 (CA)

Scott Wilson Kirkpatrick (SWK) was engaged as structural engineer and supervising officer by the MOD in relation to refurbishment of the roof at Plymouth Dockyard under GC/Works/1. Several years after the works were complete, wind lifted a large section of roof and deposited it in a nearby playing field. The contract had required 9–12 in. nails, but the contractor had used 4 in. nails. The supervising officer had been party to discussions regarding the use of the 4 in. nails, but neither he nor the contractor could remember very clearly when these discussions had taken place, or exactly what had been said. The Court of Appeal decided that the evidence was sparse and vague, and declined to find that there was any instruction under 7(1)(a) or 7(1)(m) (instructions that may be given orally), or that there had been any agreement as to the replacement. Even if the supervising officer's conduct amounted to confirmation or encouragement, this could not absolve the contractor from its duty to fix the purlins in a workmanlike manner. The MOD was therefore entitled to insist on its strict contractual rights. The Court of Appeal noted, however, that an instruction in writing was not a condition precedent to a claim by the contractor, so long as it was able to prove that the change had been agreed.

6.24   All instructions must be in writing and sent in the format and by the means agreed between the parties, which could include electronic communications (cl 1.7.2). If no means has been agreed, instructions may be sent by 'any effective means' (cl 1.7.3). It should be remembered that it may be necessary to prove that an instruction has been received and when it arrived. Clause 1.7.3 does not contain a 'deemed to have been received' provision, unlike clause 1.7.4, and therefore it is advisable to send a hard copy by recorded delivery or to record receipt of instructions at a subsequent progress meeting. Unless otherwise agreed, no special format is required for instructions, but it is often convenient to use the forms published by RIBA Publishing. An instruction in a letter would be effective, as long as the letter is quite clear. A drawing sent with a letter requiring it to be executed would constitute an instruction, but a drawing with no covering instruction may be judged to be ineffective.

6.25 Unless a special format and means have been agreed, instructions in site meeting minutes may constitute a written instruction if issued by the employer, but not if issued by the contractor, and only if the minutes are recorded as agreed at a subsequent meeting. It is possible that the instruction might not take effect until after the minutes are agreed, and it would depend on the circumstances whether the minutes were judged to be sufficiently clear to fall within the terms of the contract. It is therefore not good practice to rely on this method. The use of site instruction books should also be avoided. Although signing an instruction in a book would constitute a written instruction under the terms of the contract, there is no obligation to sign such books, and it may also be prudent not to make quick decisions on site but to wait until all the implications of the instruction can be considered.

6.26 The contractor must comply with every instruction (cl 2.1.4), provided that it is valid (i.e. provided that it is in respect of a matter regarding which the employer is empowered to issue instructions). The contractor must comply 'forthwith', which for practical purposes means as soon as is reasonably possible (cl 3.5). The requirement to comply is subject to certain exceptions:

- the contractor's consent is required to any instruction requiring a change which affects the design (cl 3.9.1), although that consent is not to be unreasonably delayed or withheld (cl 1.10);
- the contractor need not comply with a clause 5.1.2 instruction (access and use of the site, etc.) to the extent that it makes a reasonable objection (cl 3.5.1 and 3.9.2);
- if acting as principal contractor or principal designer, the contractor may have an objection under the CDM Regulations (cl 3.9.4).

6.27 Where the employer acts outside its power to give instructions as set out under the contract, the contractor is under no obligation to comply with any instruction given. If the contractor feels that an employer's instruction might not be empowered by the contract, or requires clarification, then the contractor may ask the employer to specify in writing the provisions of the contract under which the instruction is given, and the employer must do this 'forthwith' (cl 3.8). The contractor must then either comply, in which case the instruction is deemed to have been valid, or issue a notice referring the disputed instruction to the decision of an adjudicator.

6.28 Even if the contractor decides to query the instruction under clause 3.8, this does not relieve the contractor of the obligation to comply. Should the instruction be found to be valid and the contractor did not comply, it would be liable for any delay caused. If the contractor does comply, but the instruction turns out to have been invalid, the contractor may be entitled to any losses incurred through compliance.

6.29 If the contractor does not comply with a written instruction, the employer may employ and pay others to carry out the work to the extent necessary to give effect to the instruction (cl 3.6). The employer must have given written notice to the contractor requiring compliance with the instruction, and seven days must have elapsed after the contractor's receipt of the notice before the employer may bring in others. This indicates that a recorded form of delivery is desirable. Although there is no obligation to issue such notices, it would be prudent for the employer to take swift action in order to protect its interests. The employer is entitled to recover any additional costs from the contractor, i.e. the difference between what would have been paid to the contractor for the instructed work, and the costs actually incurred by the employer. These costs could include not only the carrying out of the instructed work, but also any special site provisions that need to be made, including health and safety provisions, and any additional professional fees incurred. Although it would be wise to obtain alternative estimates for all these costs wherever possible, if the work is urgent there would be no need to do so.

Changes

6.30   Employers may decide that they wish to vary the requirements at some point after the contract is signed. Under common law, neither party to a contract has the power unilaterally to alter any of its terms. Therefore, in a construction contract the employer would not have the power to make any changes unless the contract confers such a power. As some aspects of construction may be difficult to define precisely in advance, most construction contracts contain provisions allowing the employer to vary the works to some degree.

6.31   Under DB16 the employer is empowered to order specific 'Changes' to the requirements (cl 3.9.1). The term 'Change' is defined under clause 5.1 and includes alterations to the design, quality and quantity of the works, and to operational restrictions such as access to the site. The contract expressly states that no such change will vitiate the contract (cl 3.9.3), but the power does not extend to altering the nature of the contract, nor can the employer make changes after practical completion. All changes under clause 3.9 may result in an adjustment of the contract sum (cl 5.4.1) and give rise to a claim for an extension of time (cl 2.26.1) or direct loss and/or expense (cl 4.21.1).

6.32   The employer may vary the design, quality or quantity of the works; for example by requiring air conditioning in an area that had previously been naturally ventilated. The employer may add to or omit work, substitute one type of work for another or remove work already carried out (cl 5.1.1). If the change results in an alteration to the design of the works the contractor's consent must be sought (cl 3.9.1), which must not be unreasonably delayed or withheld (cl 1.10). As the contract provides for the contractor to be compensated for any increases in cost or programme, it would normally be unreasonable for the contractor to refuse consent, but it may be reasonable where the change is so extensive that it seriously disrupts the project.

6.33   The employer may vary the access to or use of the site, limitations on working space or working hours, the order in which the work is to be carried out or any restrictions already imposed (cl 5.1.2). However, the contractor need not comply with the type of change defined in clause 5.1.2 to the extent that it makes reasonable objection (cl 3.9.2). Given that the contractor will be paid for such changes it is difficult to see what might constitute a 'reasonable' objection but, for example, the instruction might make site operations almost impossible to manage. This contract provision is necessary not only to allow the employer some flexibility, but also to accommodate difficulties that may arise, for example through local authority restrictions on working hours.

6.34   Any instruction under clause 5.1 may, if it results in a suspension of work for a period longer than that stated in the contract particulars, give the contractor grounds for termination (cl 8.9.2). The consequences of such an instruction are therefore serious.

6.35   If the contractor is the principal contractor or principal designer it must, within a reasonable time of receiving an instruction effecting a change, or in respect of a provisional sum,

Table 6.4 Matters which are to be 'treated as a Change'

Clause		Notes
2.10.1	Correcting divergence between employer's requirements and definition of site boundary	Employer is required to issue an instruction which 'shall be treated as a Change'
2.12.2	Correction of inadequacies in employer's requirements	Correction 'shall be treated as a Change'
2.14.2	Discrepancies in requirements or proposals	Decision notified 'shall be treated as a Change'
2.15.2.1	Alterations of works due to statutory requirements	'shall be treated as a Change'
2.15.2.2	Change to proposals due to statutory requirements	'shall be treated as a Change' unless precluded under employer's requirements
2.15.2.3	Change to requirements due to statutory requirements	'Employer shall issue an instruction requiring a Change'
6.13.5.3	Insurance Option A: reinstatement following damage caused by terrorism	
6.13.6	Insurance Option B and C (C.2), or damage caused by an excepted risk: reinstatement following damage to the works	
Schedule 2:		
1.4.2	Termination of the named sub-contractor's employment	The contractor completes any uncompleted work, which is 'treated as a Change'
2.5	Employer's withdrawal of instruction, additional design work	'shall be treated as a Change'

inform the employer in writing if it has any objection to the instruction in relation to its obligations under the CDM Regulations (cl 3.9.4). If the contractor makes an objection, the employer may vary the terms of the instruction to remove the objection to the reasonable satisfaction of the contractor. The contractor is not obliged to comply with the instruction until the matter is resolved.

6.36　It should be noted that DB16 refers in many places to events which are to be 'treated as a Change', often without requiring the employer to issue an instruction. These are summarised in Table 6.4. The wording in each instance is not always consistent, but it appears that the contractual consequences of these deemed changes are the same. Clause 2.26.1 cites 'Changes and any other matters or instructions which under these Conditions are to be treated as, or as requiring, a Change' as a relevant event, which would cover all the matters listed below, and similar wording is included under clause 4.21.1 in relation to loss and/or expense.

Goods, materials and workmanship

6.37　Clause 2.2.2 makes it clear that all work must be carried out in accordance with the standard specified in the contract documents, or in any subsequently released contractor's design documents (which in turn must comply with the employer's requirements). The employer may appoint an architect or other professional to inspect at regular intervals to monitor the standard that is being achieved. If any changes were made in order to raise or lower the standard, then this would constitute a change. When the standard achieved appears to be unsatisfactory it can be tempting for the employer's agent (or authorised inspector) to become involved in directing the day-to-day activities of the contractor on site. Apart from being an enormous burden on the agent, such a situation could confuse the issue of who is ultimately responsible for quality and is to be avoided. The agent could, of course, draw the contractor's attention to typical areas of defective or poor quality work. Some measures for dealing with defective work are set out in the contract, as detailed below.

Defective work

6.38 The employer may instruct the contractor to open up completed work for inspection, or arrange for testing of any of the work or materials, fixed or unfixed (cl 3.12). Obviously, the employer would take this action if there were reasonable grounds for suspecting defective work or materials. No time limit is specified, but the employer should issue an instruction as soon as the need for such action becomes apparent (delay could result in escalating or unnecessary costs), although failure to ask for tests in no way relieves the contractor of the obligation to provide work according to the contract. The cost of carrying out the tests is added to the contract sum, unless it was already provided for in the employer's requirements or the contractor's proposals, or in the event that the work proves to be defective. Unless the work is defective, the contractor may also be entitled to an extension of time under clause 2.26.2 and loss and/or expense under clause 4.21.2.2.

6.39 If work is found to be defective, the employer has the power to issue an instruction 'in regard to' the removal of work, materials or goods from the site (cl 3.13.1). The employer is not required to consult with the contractor, who has no right of objection. Curiously, the employer is not given an express power to require rectification of the defective work (although such a power was given in the 1981 edition of the form). Of course, removal of the defective work would necessitate reconstruction, achieving the same result albeit in a more dramatic way, when, in fact, a small modification may be all that is needed. This would, strictly speaking, have to form the subject of a separate agreement but, in practice, the contractor is unlikely to object to an instruction that requires rectification rather than removal.

6.40 Even though instructing removal might appear excessive, considering that the contractor is already under an obligation to build the work correctly, it can be important to issue instructions that enable the clause 3.6 sanctions to be brought into operation (*Bath and North East Somerset DC* v *Mowlem*). To fall within clause 3.13.1 the instruction must specifically require removal of the work from site, however impractical. Simply drawing attention to the defective work would not be sufficient (*Holland Hannen* v *Welsh Health Technical Services*).

> *Bath and North East Somerset District Council* v *Mowlem plc* [2004] BLR 153 (CA)
>
> Mowlem plc was engaged on JCT98 (Local Authorities With Quantities) to undertake the Bath Spa project. Completion was expected to be in 2002 but work was still under way in 2003. Paint applied by Mowlem to the four pools began to peel, and the contract administrator issued Architect's Instruction no. 103 which required Mowlem to strip and repaint the affected areas.

> Mowlem refused to comply and the Council issued a notice under clause 4.1.2. Mowlem still did not comply, and the employer engaged Warings to carry out this work. Mowlem refused Warings access to the site, and the Council applied to the court for an injunction, which was granted. Mowlem appealed against the injunction, but the appeal was dismissed.
>
> Mowlem had argued that it was able to rectify all the defects and that the liquidated damages provided under the contract were the agreed remedy for delays caused. The Council was able to show that the liquidated damages were not adequate compensation for the losses suffered. Lord Justice Mance held that, in such cases, the court should examine whether the liquidated damages would provide adequate compensation and, if they would not, as in this case, it is appropriate to grant an injunction. In reaching this decision he took into account irrecoverable losses, such as the 'unquantifiable and uncompensatable damage to the Council's general public aims'.

> *Holland Hannen & Cubitts (Northern) Ltd* v *Welsh Health Technical Services Organisation* (1985) 35 BLR 1 (CA)
>
> Cubitts Ltd was employed by the Welsh Health Technical Services Organisation (WHTSO) to construct two hospitals at Rhyl and Gurnos. Percy Thomas (PTP) was the architect. Redpath Dorman Long Ltd (RDL) was the nominated sub-contractor for the design and supply of pre-cast concrete floor slabs. RDL assured WHTSO that the floors would be designed to CP 116 (concerning deflection), but the design team later required RDL to work to CP 204. Following installation, the contractor complained about extra work and costs due to adjustments to the partitions necessitated by excessive deflection of the floors, and it was established that they had been designed to CP 116 not CP 204. PTP sent three letters 'condemning' the floors, but the first did not mention clause 6(4), and none of them required removal of the work. Cubitts stopped work for 20 weeks until PTP issued instructions specifying how the defect should be resolved. Cubitts commenced proceedings, claiming compensation for delay. The claim was settled, but the relevant parties maintained their proceedings against each other for contribution. The official referee decided that RDL was liable for two-thirds of the amount paid to Cubitts and the design team for one-third. The Court of Appeal decided that this was incorrect and the correct apportionment should have been that RDL was liable for one-third and the design team for two-thirds. In reaching this conclusion it stated: 'PTP contributed very substantially to the delay which occurred, in failing to recognise the defect in the design at an earlier stage; by issuing an invalid notice in 1976, and by moving very slowly thereafter to take the necessary steps to have the defects in the flooring put right' (Robert Goff LJ).

6.41   The employer also has the power to issue such instructions requiring a change as are reasonably necessary as a consequence of an instruction in regard to defective work (cl 3.13.2). This would normally arise where part or all of the defective work is accepted, with the result that the work would fail to comply with the requirements, with possible consequential effects on other parts of the design. In such cases, as long as the instruction is 'reasonably necessary', no addition is made to the contract sum and no extension of time is given. However, the employer is required to consult with the contractor (presumably to establish the optimum solution whereby disruption and costs to the contractor are kept to a minimum), but there is no need to obtain the contractor's consent. Note, however, that there is no express provision for any deduction for accepting defective work; this would have to be negotiated between the parties. The employer should specify in writing exactly which defective work is to be retained and record the agreed deduction. Any rates and prices for that work as set out in the contract sum analysis can be used as a starting point for negotiation, but they are not the only matters to be taken into consideration (see *Oksana Mul* v *Hutton Construction Limited*).

> *Oksana Mul v Hutton Construction Limited* [2014] EWHC 1797 (TCC)
>
> This case concerned what constitutes an 'appropriate deduction' when an employer decided to accept non-conforming work. The project concerned an extension and refurbishment work to a country house using the JCT IC05 form. A practical completion certificate was issued with a long list of defects attached, and during the rectification period the employer decided to have this work corrected by other contractors. The employer then started court proceedings against the contractor, to claim back the costs of this work.
>
> A key issue was the interpretation of clause 2.30, which provides that the contract administrator can instruct the contractor not to rectify defects and 'If he does so otherwise instruct, an appropriate deduction shall be made from the Contract Sum in respect of the defects, shrinkages or other faults not made good'. In this case the contractor argued that an 'appropriate deduction' was limited to the relevant amount in the contract rates or priced schedule of works. The court disagreed. It decided that 'appropriate deduction' under clause 2.30 meant 'a deduction which is reasonable in all the circumstances', and could be calculated by any of the following: the contract rates or priced schedule of works; the cost to the contractor of remedying the defect (including the sums to be paid to third party subcontractors engaged by the contractor); the reasonable cost to the employer of engaging another contractor to remedy the defect; or the particular factual circumstances and/or expert evidence relating to each defect and/or the proposed remedial works.
>
> However, the judge also pointed out that the employer will still have to satisfy the usual principles that apply to a claim for damages, which include showing that it mitigated its loss. If the employer unreasonably refused to let the contractor rectify defects, then it is likely to find its damages limited to what it would have cost the contractor to put them right.

6.42   The employer can instruct that further tests or opening up are carried out to establish whether there are any similar problems in work already carried out (cl 3.13.3). The cost of further tests would be borne by the contractor, whether or not the additional tests proved work to be defective, provided that the instruction was reasonable. However, the contractor would have the right to an extension if the tests showed that the work was satisfactory and they caused delay. The contractor has no right of objection, and must comply with the instruction immediately.

6.43   In issuing instructions under clause 3.13.3 the employer must have due regard to the Code of Practice included in Schedule 4 of the form. There is no requirement to adhere to the code exactly but, as its observance may be considered evidence that the instructions were reasonable, the employer would be wise to pay it close attention.

6.44   Clause 2.1.1 requires the contractor to carry out the work 'in a proper and workmanlike manner' and in accordance with the construction phase plan. Where there is any failure to do so, clause 3.14 entitles the employer to 'issue such instructions (whether requiring a Change or otherwise) as are in consequence reasonably necessary'. To the extent that such an instruction is necessary, nothing is added to the contract sum and no extension of time given. This is similar to the employer's power under clause 3.13.2, discussed at paragraph 6.41.

Sub-contracted work

6.45   Under clause 3.3.1 the contractor may only sub-contract work with the written consent of the employer. Failure to comply with this restriction would be a default, providing grounds for termination under clause 8.4.1.4. Under clause 1.10, however, the employer's permission cannot be unreasonably delayed or withheld. It is suggested that permission is required for each instance of sub-contracting, rather than agreeing to sub-contracting in principle. Any permission is stated not to affect the contractor's obligations under the contract (cl 3.3.1), for example to carry out the work to the required standard or to meet the completion date.

6.46   The form states 'where considered appropriate' the contractor shall engage the sub-contractor on the JCT Design and Build Sub-Contract (cl 3.4). Although not an absolute requirement, the contractor ought to consider using this form for all sub-contracting, unless there are sensible reasons why this is not appropriate. Whatever form of sub-contract is used, however, it must include certain conditions, and clause 3.4 states that the sub-contract must provide that:

- the sub-contract is terminated immediately on termination of the main contract (cl 3.4.1);
- unfixed materials and goods placed on the site by the sub-contractor (the 'Site Materials') shall not be removed without the written consent of the contractor (cl 3.4.2.1);
- it shall be accepted that materials or goods included in an interim payment by the employer become the property of the employer (cl 3.4.2.1.1);
- it shall be accepted that any materials or goods paid for by the main contractor prior to being included in a payment by the employer become the property of the main contractor (cl 3.4.2.1.2);
- that each party undertakes to comply with the CDM Regulations (cl 3.4.2. 3);
- the sub-contractor has a right to interest on late payments by the contractor at the same rate as that due on main contract payments (cl 3.4.2.4).

6.47   In addition, clause 3.4 sets out requirements where third party rights and/or warranties are to be granted by the sub-contractor (cl 3.4.2.5). For example, that warranties will be executed within 14 days of receipt of a notice from the contractor (and as a deed, if applicable) and that third party rights will be vested by the sub-contractor upon notice by the contractor. Finally (and very important in the context of design and build), if documents, information or licences may be required from the sub-contractor in relation to as-built drawings, the CDM Regulations and/or the BIM protocol, the sub-contract must make provision for this (cl 3.4.3).

6.48   Most of these terms protect the position of the employer and the provisions regarding unfixed goods and materials are of particular importance in this respect. If a main contractor sub-contracts on other terms, with resultant losses to the employer, then the contractor may be liable for breach of contract.

Named sub-contractors

6.49   DB16 includes provisions for naming sub-contractors, which can be very useful if the employer wishes to involve particular firms in certain aspects of design and construction (the resulting risk distribution is summarised in Table 6.5). The provisions for naming of sub-contractors are set out in Schedule 2, Supplemental Provision 1 and will only apply if

Table 6.5 Distribution of risk between parties when named sub-contractors are used

	Revision to contract sum?	Extension of time?	Loss and/or expense?
Named sub-contractor's progress causes delay	no	no	no
Change instructions dealing with named sub-contractor problems (e.g. to 'remove the grounds') (Schedule 2:1.1.2)	yes (5.2)	yes (2.26.1)	yes (4.21.1)
Delay in issuing instructions dealing with named sub-contractor problems	no	yes (2.26.6)	yes (4.21.5)
Following termination, completion of work by contractor under Schedule 2:2.1.5			
If termination due to default of contractor, or if contractor fails to comply with Schedule 2:1.1.3 (notification and consent)	no	no	no
If termination not due to default of contractor and contractor complies with Schedule 2:1.1.3	yes	yes	yes

incorporated under the contract particulars. Supplemental Provision 1.1 states that its provisions apply 'Where the Employer's Requirements state that work ('Named Sub-Contract Work') is to be executed by a named person as the Contractor's sub-contractor (a 'Named Sub-Contractor')'. It appears from the following clauses that the requirements must give the actual name of the person appointed to carry out the work, not simply state that this work is to be carried out by a person to be named by the employer.

6.50 There is no indication as to what additional information should be included in the requirements at tender stage. This is due to the wide range of information that might be available, which is dependent on how fully developed the design is at that stage. It would be sensible to give as much information as possible including, if available:

- full details of the company;
- the work to be carried out;
- price, terms and conditions, including any warranty required;
- design drawings, etc. already prepared;
- the price of the work and tender details if obtained.

6.51 In order to collate this information, the employer must first have invited sub-contract tenders. If this has not happened, the employer will provide only limited information to identify the named sub-contractor and the related work, and the contractor will then proceed to obtain a tender.

6.52 The contractor must enter into an agreement with the named person 'as soon as reasonably practicable' after entering into the main contract with the employer (Schedule 2:1.1.1). The sub-contractor should be given Supplemental Provision 1 and, in particular, notification of paragraph 1.5 (which protects the employer in the event of termination) must be incorporated. The contractor should engage the sub-contractor on the JCT Design and Build Sub-Contract, or comply with clause 3.4 (see paragraph 6.46). Other than this, no particular terms are required for the sub-contract, unless the employer has stipulated conditions in the requirements.

6.53 If the contractor is unable to enter into a sub-contract in accordance with the particulars in the main contract documents, the contractor must immediately inform the employer of the reasons that have prevented this from happening (Schedule 2:1.1.2). Provided the contractor has acted reasonably, the employer must then by a change instruction either remove the reason, or direct the contractor to undertake the work itself (or have it carried out by a sub-contractor selected by the contractor and approved by the employer), or omit by a change the named sub-contract work. Whatever option is selected, the change must be valued under clause 5.2, may give rise to a claim for an extension of time and will also be a matter with respect to a claim for direct loss and/or expense (Schedule 2:1.2). If the instruction removes the named sub-contractor, it may not name another (Schedule 2:1.1.2). However, if the work is omitted, the employer could arrange for it to be carried out by a person employed directly by the employer under clause 2.6.

6.54 If the named sub-contractor's employment is terminated by the contractor, the contractor is required to carry out any outstanding work (Schedule 2:1.4). This could be sub-contracted with the consent of the employer. The completion of the work is treated as a change, except if the termination resulted from the contractor's default (paragraph 10.34 discusses termination in more detail).

Work not forming part of the contract/persons engaged by the employer

6.55 Under clause 2.6, the employer may engage persons directly to carry out work that does not form part of the contract while the main contractor is still in possession. This may include statutory undertakers when employed by the employer, but not where they are carrying out the work in pursuance of their statutory duties. If the contract documents have included 'the information necessary to allow the Contractor to carry out and complete the Works or each relevant Section in accordance with this Contract', then the contractor must permit the employer to execute such work. Otherwise, the employer can only do this with the contractor's consent. The consent may not be unreasonably delayed or withheld.

6.56 Clauses 6.1 and 6.2 make it clear that the contractor is not liable for, nor is it required to indemnify the employer against, claims due to injury or damage to property caused by the directly engaged person. The employer should therefore ensure that insurance cover is arranged in respect of any act or neglect on the part of the persons to be employed. The employer should also take note that any disruption to the contractor's working could lead to an extension of time (cl 2.26.6), to a claim for loss and expense (cl 4.21.5), or even to the contractor terminating the contract (cl 8.9.2). The employer is therefore at considerable risk, and should avoid this route if at all possible.

Making good defects

6.57 The contractor is required to make good any 'defects, shrinkages or other faults' which appear and are notified by the employer to the contractor (cl 2.35). The defects are limited to those that result from 'failure of the Contractor to comply with his obligations'. This would not, for example, include shrinkages or general wear and tear due to occupation, which would be expected even if the work had been carried out as specified.

6.58   The notification should take the form of a schedule, which must be issued to the contractor not later than 14 days after the end of the rectification period (cl 2.35.1). The obligation appears to be limited to those defects that appear after practical completion (latent defects) and does not cover defects that can be identified at practical completion (patent defects). Nevertheless, it is common practice to include any minor items that were outstanding at that stage, as this is a practicable way of dealing with such matters.

6.59   However, it is important to note that the obligation to make good appears to be limited to those defects notified by the employer. Although the contractor may be liable for defects which are not notified, the employer would be unlikely to be able to claim the full cost of repair of any defect made good at a later date. It is therefore important that the employer prepares a comprehensive schedule (*Pearce and High* v *Baxter*, see also *Oksana Mul* v *Hutton* at paragraph 6.41). The employer is required to issue the notification not later than 14 days after the end of the rectification period, the only point at which the contract requires the employer to issue such a schedule. The employer may instruct the contractor to make good a defect at an earlier stage, but this would normally only be used for serious and urgent problems (cl 2.35.2).

> *Pearce and High* v *John P Baxter and Mrs A S Baxter* [1999] BLR 101 (CA)
>
> The Baxters employed Pearce and High on MW80 to carry out certain works at their home in Farringdon. Following practical completion, the architect issued interim certificate no. 5, which the employer did not pay. The contractor commenced proceedings in Oxford County Court, claiming payment of that certificate and additional sums. The employer, in its defence and counterclaim, relied on various defects in the work that had been carried out. Although the defects liability period had by that time expired, neither the architect nor the employer had notified the contractor of the defects. The Recorder held that clause 2.5 was a condition precedent to the recovery of damages by the employer, and further stated that it was a condition precedent that the building owner should have notified the contractor of patent defects within the defects liability period. The employer appealed and the appeal was allowed. Lord Justice Evans stated that there were no clear express provisions within the contract which prevented the employer from bringing a claim for defective work, regardless of whether notification had been given. He went on to state, however, that the contractor would not be liable for the full cost to the employer of remedying the defects if the contractor had been effectively denied the right to return and remedy the defects itself.

6.60   If the employer decides to accept any defective work, then this should be clearly confirmed in writing. The full extent of the problem should be carefully established before such a course of action is taken, and an appropriate deduction from the contract sum agreed, as it is unlikely that the employer would thereafter be able to claim for consequential problems or further remedial work.

6.61   Once satisfied that all the notified defects have been made good, the employer must issue a notice to that effect (the 'Notice of Completion of Making Good', cl 2.36). The notice must not be unreasonably delayed or withheld. The notice is one of the preconditions for the final payment.

6.62   The contract does not state what should happen in respect of defects which appear after the issue of the schedule of defects, or after the end of the rectification period but before the final payment. It is, however, clear from clause 2.35 that the employer no longer has the power to instruct that these are made good. It is suggested that in such circumstances there would be two possible courses of action. The first would be to make an agreement with the contractor to rectify the defects before the final payment is made. If the contractor refused to do this, an amount could be deducted from the contract sum to cover the cost of making good the work, but this would involve some risk to the employer. The second and less risky course would be to have the defective work made good by another contractor, and deduct the cost from the contract sum. This would involve a delay to the final payment and would probably be disputed by the contractor.

6.63   The contractor's liability for defective work is not limited to defects notified by the employer, and does not end with the final payment. The contractor is still liable for losses suffered, but no longer has the right to return to site to correct defective work. The employer's remedy is to bring an action at common law. The rectification period is therefore a sensible procedure which benefits the parties in affording an opportunity to remedy problems at a reasonable cost, without the problems associated with bringing a legal action.

7 Sums properly due

7.1 DB16 is a lump sum contract, which means that all the work described in the contract documents, including completion of the design to meet the employer's requirements, is to be carried out for the agreed sum, and there is no provision for any remeasurement. The contract sum, which is the tender figure accepted or agreed following negotiation, is entered in Article 2.

7.2 Although the employer may assume that the contract sum is 'fixed', in design and build procurement this is rarely the case in practice. Most forms of contract include provisions to deal with situations where the employer requires changes to the design, usually with heavy financial penalties. Additional provisions may be introduced to accommodate changing circumstances, such as disruption to the works or changes in statutory charges, which serve to reduce the considerable risk to the contractor of tendering a truly 'fixed' price and which would normally result in a high tender figure.

7.3 DB16 includes several provisions which provide for adjustment to the amount to be paid and refers, in Article 2, to the contract sum 'or such other sum as becomes payable'. The provisions for adjustment are summarised in clause 4.2. It should be noted that the contract sum itself does not change and the provisions refer to adding or subtracting amounts from the contract sum to reach a revised figure.

7.4 There are many reasons why the amount finally payable may differ from the contract sum (Table 7.1). If the employer instructs a change to the employer's requirements, then the

Table 7.1 Adjustments to the contract sum

Clause	
2.10.1	Correcting error in definition of site boundary
2.12.2	Correction of inadequacies in employer's requirements
2.14.2	Discrepancies within the employer's requirements
2.15.2.1	Alterations of works due to statutory requirements
2.15.2.2	Change to proposals due to statutory requirements
2.15.2.2	Change to requirements due to statutory requirements
2.18	Statutory fees and charges, if stated to be part of a provisional sum
2.20.1	Infringement of patent rights
2.35	Employer's acceptance of defective work
3.12	Costs of tests
3.15.2	Instructions relating to antiquities on site

Table 7.1 Adjustments to the contract sum – Continued		
4.11.2	Costs due to suspension	
4.21	Loss and/or expense	
5.3	Changes to the employer's requirements, or imposition by the employer of obligations	
6.11.3	Net additional costs to contractor of terrorism cover	
6.12.2	Costs due to failure to take out insurance	
6.13.5.3	Insurance Option A: reinstatement following damage caused by terrorism	
6.13.6	Insurance Option B and C (C.2) or damage caused by an excepted risk: reinstatement following damage to the works	
Schedule 2:		
1.1.2	Change instructions to deal with named sub-contractor problems	
1.4.2	completion work by contractor following termination of the named sub-contractor's employment	
2.5	Additional design work prior to employer withdrawing instruction	
3.5	Accepted or agreed estimates of direct loss and/or expense	
4.3	Accepted acceleration quotation	
7.3	Accepted or agreed cost savings proposed by the contractor	
Schedule 7	Fluctuations options	

amount payable will be adjusted accordingly. There is also the possibility of claims from the contractor for loss and expense arising from intervening events which could not be foreseen at the time of tendering. Under the supplemental provisions, the contractor is encouraged to propose cost-saving and value-improvement measures, which may also result in a change (Supplemental Provision 7) and, if an amount is agreed following an acceleration quotation, then this is to be added to the contract sum. Fees or charges in respect of statutory matters for which no provision is made in the contract documents will require an adjustment to the contract sum. DB16, like most contracts, contains 'fluctuations' provisions allowing for adjustments in the event of changes in statutory charges or the market price of materials and labour. VAT is, of course, not included in the contract sum.

7.5 There will, therefore, almost inevitably need to be adjustments during the course of the contract. However, DB16 clause 4.1 makes it clear that the only alterations that may be made are those provided for in the terms. Any such amounts should be added or deducted as soon as they have been ascertained in whole or in part and included in the next interim payment (cl 4.3). There is no allowance for adjustments due to arithmetical errors in pricing made by the contractor.

Valuation of changes in the employer's requirements and provisional sum work

7.6 There are three mechanisms by which a change can be valued under the provisions of the contract. Clause 5.2 requires that all changes and all instructions relating to the expenditure of provisional sums are, unless agreed between the parties, valued using the 'Valuation Rules' in accordance with the provisions of clauses 5.4 to 5.7 (see paragraphs 7.14 to 7.16). The valuation would be made by the employer, although there would be nothing to prevent the contractor preparing an assessment. If the supplemental provisions are incorporated, the second mechanism is through the submission of contractor's estimates (Schedule 2:2). In both cases, the valuation should include a sum in respect of any additional design work required by the instruction. Under Supplemental Provisions 7 and 8, the contractor is encouraged to propose changes (cost-saving and value-improvement measures, and environmental performance improvement measures). The value of any resultant change will be a matter for agreement between the parties.

Supplemental Provision 2: contractor's estimates

7.7 If Supplemental Provision 2 is incorporated, the contractor must, on receiving an instruction requiring a change, submit an estimate to the employer, unless the employer states otherwise in the instruction or informs the contractor within 14 days of the date of the instruction that such an estimate is not required, or unless the contractor makes reasonable objection to the provision of the estimate (Schedule 2:2.2). The estimate must be submitted to the employer within 14 days of the date of the instruction, or within any other period agreed or 'as may be reasonable in all the circumstances'.

7.8 The estimate is to comprise (Schedule 2:2.3):

- the value of the change (with supporting calculations referring to the contract sum analysis);
- the additional resources required to comply;
- a method statement for compliance;
- the length of any extension of time required;
- the amount of any loss and/or expense required.

7.9 The contractor and employer are then required to 'take all reasonable steps' to agree those estimates. If they can agree, the agreement is binding on both parties and should be put in writing, but if they cannot agree, the employer can either instruct the change, which could be valued under the valuation rules (see below) or withdraw the instruction. In the second case, the contractor will be paid for any design work carried out in order to submit the estimate. Although the clauses do not say so, it would be possible to agree on only part of the estimate (e.g. the price) and determine the other parts using the normal procedures.

7.10 These provisions differ from the Schedule 2 provisions in SBC16 in that the estimate is automatic unless the employer indicates otherwise and the contractor is not paid for the cost of any estimate of a change which the employer does not implement, except that it would be paid for any design work (Schedule 2:2.5).

7.11 If the contractor fails to provide an estimate as required, the instruction will be dealt with under the provisions of clauses 2.23 to 2.26, 3.9 and 4.20, except that any addition to the contract sum will not be included until the final statement is prepared, and the contractor will not be entitled to any financing charges as a result of compliance with the instruction.

Cost saving and value improvement

7.12 If Supplemental Provision 7 is incorporated, the contractor is encouraged to propose cost-saving and value-improvement measures. Any proposed measure should relate to the design and specification and/or to the programme, and should result in an immediate saving or a saving in the life-cycle costs of the project. Once the contractor has submitted a proposal with relevant details, the parties are required to negotiate with a view to agreeing its value and, if the negotiations are successful, the change and the cost saving are confirmed in an instruction.

7.13 Supplemental Provision 8 relates to environmental considerations. The contractor is encouraged to propose amendments to the works which, if implemented, would result in an improvement in environmental performance. The proposed measures are required to be 'economically viable', which could mean a saving (possibly in life-cycle costs) or could result in an addition to the contract sum. If agreed, the measures are instructed as a change.

Valuation under the valuation rules

Measurable work – contract sum analysis

7.14 Omissions are valued in accordance with the values in the contract sum analysis (cl 5.4.3). Additional work of 'similar character' to that in the contract sum analysis is valued according to prices stated in that document, even if those prices contain an error (*Henry Boot Construction Ltd* v *Alstom Combined Cycles*), with a fair allowance being made if the conditions change or the quantity changes significantly (cl 5.4.2). An example of dissimilar conditions might include the instructed work being carried out in winter, whereas under the contract sum analysis it had been assumed that it would be carried out in summer. Such an assumption, however, would have to be clear from an objective analysis of the contract documents (*Wates Construction* v *Bredero Fleet*).

> *Henry Boot Construction Ltd* v *Alstom Combined Cycles* [2000] BLR 247
>
> By a contract formed in 1994, Alstom Combined Cycles employed Henry Boot to carry out civil engineering works in connection with a combined cycle gas turbine power station for PowerGen plc at Connah's Quay in Clwyd. During post-tender negotiations, a price of £258,850 was agreed for temporary sheet piling to trench excavations. Disputes arose regarding the valuation of this work, and these disputes were initially taken to arbitration. The arbitrator found that the agreed figure contained errors that effectively benefited Boot. Boot argued that, nevertheless, the figure should be used to value the work under clause 52(1). The arbitrator decided that clause 52(1)(a) and (b) were inapplicable, and that 52(2) should be applied to achieve a fair valuation. Boot appealed to the Technology and Construction Court, and Judge Humphrey Lloyd decided that the mistake made no difference; the agreed rate should be used even if the results were unreasonable. Clause 52(2) created only a limited exception where the scale or nature of the variation itself made it unreasonable to use the contract rates.

> *Wates Construction (South) Ltd* v *Bredero Fleet Ltd* (1993) 63 BLR 128
>
> Wates Construction entered into a contract on JCT80 to build a shopping centre for Bredero. Some sub-structural work differed from that shown on the drawings and disputes arose regarding the valuation of the works, which were taken to arbitration. In establishing the conditions under which, according to the contract, it had been assumed that the work would be carried out, the arbitrator took into account pre-tender negotiations and the actual knowledge that Wates gained as a result of the negotiations, including proposals that had been put forward at that time. Wates appealed and the court found that the arbitrator had erred by taking this extrinsic information into consideration. The conditions under which the works were to be executed had to be derived from the express provisions of the bills, drawings and other contract documents.

Daywork – fair valuation

7.15　When considering work which is not of similar character to that in the contract sum analysis, the contract states that 'a fair valuation shall be made' (cl 5.4.2). Where the appropriate basis of a fair valuation is daywork, clause 5.5 sets out the rules to be followed in assessing the amount. Otherwise, it would, in the first instance, be up to the employer to determine. If the contractor and employer cannot reach agreement on a fair rate, and the dispute cannot be resolved amicably, then the matter must be resolved in adjudication.

7.16　Any valuation of omitted, additional or substituted work should make appropriate adjustment for the provision of site administration and temporary works (cl 5.4.4).

Reimbursement of direct loss and/or expense

7.17　The objective of clauses 4.19 to 4.23 is to enable the contractor to be reimbursed for direct loss and/or direct expense suffered as a result of delay or disruption, and for which the contractor is not reimbursed under any other provision in the contract. Alternatively, the contractor may be able to claim for general damages for breach of contract at common law where the delay etc. is caused by a breach of contract, but this would need to be pursued through adjudication, arbitration or litigation (cl 4.23).

7.18　The contractor is entitled to be reimbursed for loss and/or expense incurred as a result of any occurrence of a 'Relevant Matter' set out in clause 4.21. The amount to be paid is determined under the procedure in clause 4.20 or under Supplemental Provisions 2 or 3 (Schedule 2), if incorporated (Figure 7.1).

7.19　Under clause 4.19, the employer is only obliged to compensate for direct loss and/or expense where the contractor has complied with the procedure set out in clause 4.20 (cl 4.19.1). This requires the contractor to make a written application, and to submit it promptly, i.e. 'as soon as the likely effect of the Relevant Matter on regular progress or the likely nature and extent of any loss and/or expense arising from a deferment of possession becomes (or should have become) reasonably apparent to him' (cl 4.20.1). The notice is to be accompanied by, or followed by, an assessment of the losses already incurred and those likely to be incurred (cl 4.20.2). The contractor must keep the employer updated at monthly intervals until all information reasonably required and necessary for ascertaining the amount due has been supplied to the employer (cl 4.20.3).

Figure 7.1 Ascertainment of loss and/or expense (L/E)

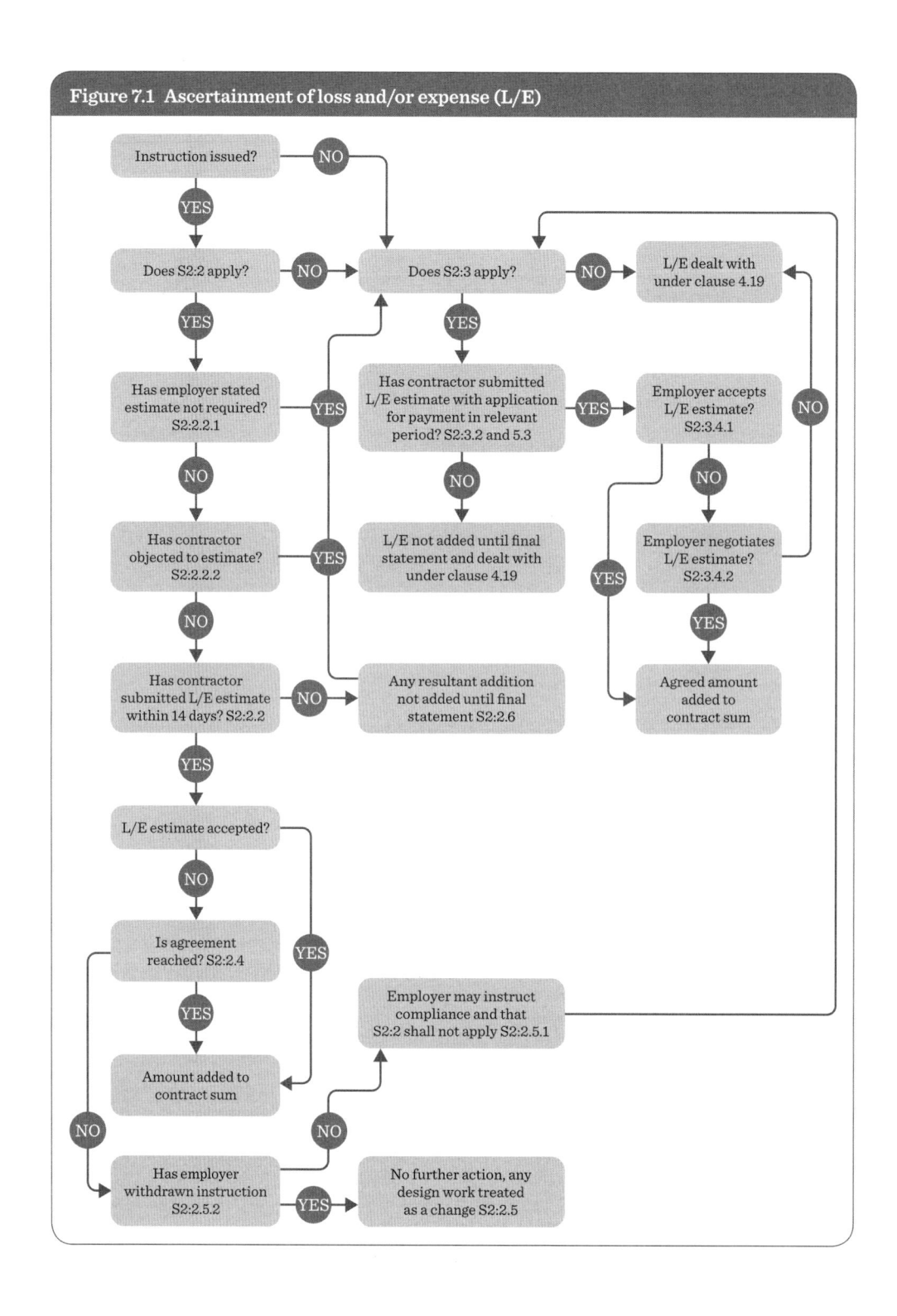

7.20 The employer is required to notify the contractor of the ascertained amount of loss and/or expense within 28 days of receipt of the initial assessment and information, and subsequently within 14 days of receipt of each monthly update of the assessment and information (cl 4.20.4). Each ascertainment must be made by reference to the information supplied by the contractor and be in sufficient detail to allow the contractor to identify differences between its own assessment and the employer's ascertainment.

7.21 The procedure is more detailed and contains stricter time limits than that in DB11. It ensures that the employer is kept fully up to date with the effect and likely costs associated with any relevant event. As well as allowing the employer to budget for the additional costs, there may be steps that the employer can take at an early stage to minimise the potential increase. It also ensures that the contractor is informed at an early date of any disagreement by the employer with the contractor's assessment, and is updated on a regular basis as to any changes in that position.

7.22 Importantly, clause 4.19, as well as stating the contractor's right to loss and expense, also states that the entitlement is 'subject to compliance with the provisions of clause 4.20'. The courts held even on earlier, less clear versions of this clause that the right to loss and/or expense could be lost if the contractor did not act promptly (see *London Borough of Merton v Leach*). Given the new wording, there is no doubt that the employer could refuse to consider late applications. However, as it is still arguable that in some circumstances the contractor might retain the right to claim this amount under common law (a right confirmed by clause 4.23), it may be sensible to agree that it should be dealt with under the contract, in cases where the procedural failing on the part of the contractor is minor.

> *London Borough of Merton* v *Stanley Hugh Leach Ltd* (1985) 32 BLR 51 (ChD)
>
> Stanley Hugh Leach entered into a contract on JCT63 with the London Borough of Merton to construct 287 dwellings. The contract was substantially delayed and a dispute arose regarding this delay and related claims for loss and expense. The dispute went to arbitration and the arbitrator made an interim award on a number of matters. The local authority appealed and the court considered 15 questions framed as preliminary issues. Among other things, the court stated that applications for direct loss and/or expense must be made in sufficient detail to enable the architect to form an opinion as to whether there is, in fact, any loss and/or expense to be ascertained. If there is, then it is the responsibility of the architect to obtain enough information to reach a decision. This responsibility could, of course, include requests for information from the contractor. The court also held that the application must be made within a reasonable time and not so late that the architect was no longer able to form an opinion on matters relevant to the application.

Alternative procedure using Supplemental Provision 2 estimate

7.23 Supplemental Provision 2 (Schedule 2), if incorporated, provides an alternative means of determining the loss and/or expense incurred as a result of a change instruction. Under this provision, if a change is required under an instruction, the contractor must submit an estimate of any consequential loss and/or expense *before* carrying out the instruction. If the estimate is accepted, then this becomes a binding agreement as to the amount to be paid in respect of that instruction, and no further claims can be made irrespective of the extent of losses suffered by the contractor. If the estimate cannot be agreed, then either the work can be instructed or the instruction withdrawn.

Alternative procedure using Supplemental Provision 3 estimate

7.24 Supplemental Provision 3 (Schedule 2) provides an alternative means of determining the amount of loss and/or expense to be paid as a result of any of the relevant matters cited in clause 4.21. If Supplemental Provision 3 is incorporated, then with each application for an interim payment the contractor must submit an assessment of the loss and/or expense incurred during the preceding period. For example, where an application for payment for September is made in early October, it should include the amount of direct loss and/or expense suffered in August. Any loss and/or expense which has been, or is being, dealt with under Supplemental Provision 2 should not be included in the application, as a binding agreement on this will have already been reached.

7.25 The employer must, within 21 days of the estimate (Schedule 2:3.4), give notice to the contractor that it either accepts the estimate, or that it wishes to negotiate an agreement on the amount (and, if no agreement is reached, refer the matter to arbitration or legal proceedings), or that the provisions of clause 4.20 as set out above will apply. During these 21 days, the employer may request further information, but may not use this as a means of delaying a decision.

7.26 If the estimate is accepted or agreed, the amount is added to the contract sum and no further amounts can be claimed for losses during that period. In other words, the loss and/or expense are finally determined for that period, in contrast to the clause 4.20 provisions, where the amount could be re-assessed following the provision of further information from the contractor. If the amount cannot be agreed, then the employer can either waive the requirement in Supplemental Provision 3.2, so that the matter can be dealt with under clause 4.20, or refer the matter to adjudication.

7.27 For subsequent payments, the contractor must submit estimates of any additional loss and/or expense incurred during the immediately preceding period. Each period is treated discretely, and the contractor cannot, in any application, revise any amount claimed in respect of a previous period.

7.28 If the contractor fails to submit an estimate as required by the provisions, then the ascertainment is dealt with under clause 4.20, and is not added to the contract sum until the final payment. When this applies, the contractor may not claim financing charges for the period between when the loss was incurred and the final payment. The provisions in Supplemental Provision 3 provide a sensible method of resolving the issues of loss and/or expense as the project proceeds, although the penalties on the contractor for non-compliance are quite severe (Schedule 2:3.6).

Matters for which loss and expense can be claimed

7.29 Claims under clause 4.19 can only be made for loss and expense suffered through deferment of possession, or the relevant matters listed in clause 4.21 which, it should be noted, include delay in the receipt of planning and other approvals. Other losses are irrecoverable under the contract, although disputed claims may be referred to adjudication, arbitration or litigation. The matters listed in clause 4.21 are concerned with situations where the loss or expense is attributable to the employer, including 'any impediment, prevention or default, whether by act or omission, by the Employer', and excluding the neutral causes which feature in the extension of time provisions of clause 2.26. It should be noted that costs and expenses resulting from the contractor exercising its right to suspend work under clause 4.11 are not dealt with under clause 4.20, but treated separately (see paragraph 8.37).

7.30   Clause 4.19 refers to regular progress of the works being 'materially affected' by the relevant event. The phrase 'regular progress of the Works or any part of them has been or is likely to be materially affected' allows a claim for disruption not anticipated when tendering. This could include situations where an overall delay to the programme is experienced (often termed 'prolongation') for which an extension of time may have been awarded. However, it can also include disruption to the planned sequence that does not cause any overall delay, provided it can be shown that losses were suffered as a result. Any disruption claim should be related to the progress necessary to complete the works by the completion date, not necessarily the actual sequences of events on site, and the disruption would have to be significant for it to entitle the contractor to compensation.

7.31   In ascertaining the loss and expense, the contractor or employer must determine what has actually been suffered. The sums that can be awarded may include any loss or expense that has arisen directly as the result of the 'matter'. The loss and expense award is, in effect, an award of damages. Its assessment should be made on the same principles as those a court would adopt when awarding damages for breach of contract. In broad terms, the object of the award is to return the contractor to the position in which it would have been but for the disturbance. The contractor should be able to show that it has taken reasonable steps to mitigate its loss, and the losses must have been reasonably foreseeable as likely to result from the 'matter' when the contract was entered into.

7.32   The following are items which may be included in the award:

- increased preliminaries;
- overheads;
- loss of profit;
- uneconomic working;
- increases due to inflation;
- interest or finance charges.

7.33   The items claimed must not be recoverable by the contractor under any other term of the contract (and, for example, duplication of a claim under clause 5.4 must be avoided). Prolongation costs, such as on-site overheads, would normally only be claimable for periods following the original completion date. (For head office overheads, etc., see *McAlpine* v *Property and Land Contractors* below.) Interest may also be recoverable, but only if it can be proved to have been a genuine loss (*F G Minter* v *WHTSO*).

> *Alfred McAlpine Homes North Ltd* v *Property and Land Contractors Ltd* (1995) 76 BLR 59
>
> An appeal arose on a question of law arising out of an arbitrator's award regarding the basis for awarding direct loss and expense with respect to additional overheads and hire of small plant, following an instruction to postpone the works. The judgment contains useful guidance on the basis for awarding direct loss and expense. To 'ascertain' means to 'find out for certain'. It is not

necessary to differentiate between 'loss' or 'expense' in a head of claim. Regarding overheads, a contractor would normally be entitled to recover as a 'loss' the shortfall in the contribution that the volume of work had been expected to make to the fixed head office overheads, but which, because of a reduction in volume and revenue caused by the prolongation, was not in fact realised. The fact that 'Emden' or 'Hudson' formulae depend on certain assumptions means that they are frequently inappropriate. The losses on the plant should be the true cost to the contractor, not based on notional or assumed hire charges.

F G Minter Ltd v *Welsh Health Technical Services Organisation* (1980) 13 BLR 1 (CA)

Minter was employed by Welsh Health Technical Services Organisation (WHTSO) under JCT63 to construct the University Hospital of Wales (second phase) Teaching Hospital. During the course of the contract, several variations were made and the progress of the works was impeded by the lack of necessary drawings and information. The contractor was paid sums in respect of direct loss and/or expense, but the amounts paid were challenged as insufficient. The amounts had not been certified and paid until long after the losses had been incurred, therefore the figures should have included an allowance in respect of finance charges or interest. Following arbitration, several questions were put to the High Court, including whether Minter was entitled to finance charges in respect of any of the following periods:

(a) between the loss and/or expense being incurred and the making of a written application for the same;
(b) during the ascertainment of the amount; and/or
(c) between the time of such ascertainment and the issue of the certificate including the ascertained amount.

The court answered 'no' to all three questions and Minter appealed. The Court of Appeal ruled that the answer was 'yes' to the first question and 'no' to the others.

7.34 The contractor would normally provide full details and particulars of all items concerned with the alleged loss and/or expense. These should identify which of the losses claimed relate to each of the 'matters' that have occurred. This is sometimes compromised by the use of a 'rolled up' or composite claim approach, where it is not really practicable to separate and itemise the effects of a number of causes. This has been accepted by the courts, provided that as much detail as possible has been given, and provided that all disturbance was due to matters under clause 4.21 and not caused by the contractor.

7.35 Formulae such as the 'Hudson' or 'Emden' formulae are sometimes used to estimate head office overheads and profit, which may be difficult to substantiate. Such formulae can be used only where it has been established that there has been a loss of this nature. To do this, the contractor must be able to show that, but for the delay, the contractor would have been able to earn the amounts claimed on another contract, for example by producing evidence such as invitations to tender which were declined. Such formulae may be useful where it is difficult to quantify the amount of the alleged loss, provided that a check is made that the assumptions on which the formula is based are appropriate.

7.36 Although direct loss and/or expense are matters of money and not time, which are quite separate issues, there is often a practical correlation in the case of prolongation. Any general implication that there is a link would be incorrect and, in principle, disruption claims and claims for delay to progress are independent. An extension of time, for example, is not a condition precedent to the award of direct loss and/or expense (*H Fairweather & Co.* v *Wandsworth*, see paragraph 5.33).

7.37 Applications by or claims from the contractor made under the contract must be dealt with according to the procedures contained in the contract. Failure correctly to ascertain an amount properly due could leave the employer liable in damages for breach of contract (*Croudace* v *London Borough of Lambeth*).

> *Croudace Ltd* v *London Borough of Lambeth* (1986) 33 BLR 20 (CA)
>
> Croudace entered into an agreement with the London Borough of Lambeth to erect 148 dwelling houses, some shops and a hall. The contract was on JCT63 and the architect was Lambeth's chief architect and the quantity surveyor was its chief quantity surveyor. The architect delegated his duties to a private firm of architects. Croudace alleged that there had been delays and that they had suffered direct loss and/or expense and sent letters detailing the matters to the architects. In reply, the architects told Croudace that they had been instructed by Lambeth that all payments relating to 'loss and expense' had to be approved by the borough. The chief architect then retired and was not immediately replaced. There were considerable delays pending the appointment of a successor and Croudace began legal proceedings. The High Court found that Lambeth was in breach of contract in failing to take the necessary steps to ensure that the claim was dealt with, and was liable to Croudace for this breach. The Court of Appeal upheld this finding.

Fluctuations

7.38 In some projects it may be advantageous to insist on a 'fixed' or 'guaranteed' price, whereby the contractor accepts the risk of all changes in the cost of the works due to statutory revisions and market price fluctuations. However, pricing with this degree of certainty will, in some economic climates, result in higher tender figures, as the contractor will need to allow for possible increases, particularly if the contract period is relatively lengthy. In order to avoid inflated tenders, most contracts allow for some 'fluctuations', whereby the employer accepts some of these risks.

7.39 In DB16 the default fluctuations provisions are set out in Schedule 7, which allows for contribution, levy and tax fluctuations (Option A). The traditional full fluctuations in labour and materials (Option B) and the use of price adjustment formulae (Option C) are no longer included in the contract, but are available from the JCT website, and are referred to in the form. The contract particulars (cl. 4.2, 4.12 and 4.13) set out the three options; if no selection is made then Option A (i.e. so-called 'fixed price') applies.

7.40 Option A provides for full recovery of all fluctuations in the rates of contributions, levies and taxes on the employment of labour, and in the rates of duties and taxes on the procurement of materials. In short, the only amounts payable are those arising out of an Act of Parliament or delegated legislation. However, contractors have pointed out that many less obvious increases are not included, therefore a 'percentage addition' is made to allow for these. The agreed percentage is entered in the contract particulars. Option C allows for adjustment based on the use of formulae: it does not necessarily take account of the actual costs, but is relatively simple to operate, and is generally considered by contractors to be a fair adjustment.

7.41　Where a contract includes recovery for fluctuations, they will, in the absence of anything to the contrary, be payable for the whole time the contractor is on site even though it fails to complete within the contract period (*Peak Construction* v *McKinney Foundations*). There is a so-called 'freezing' provision in DB16 (paragraphs A.9, B.10 and C.6), but this depends on the text of clauses 2.23 to 2.26 being left unamended and all notices of delay being properly dealt with by the employer.

> *Peak Construction (Liverpool) Ltd* v *McKinney Foundations Ltd* (1970) 1 BLR 111 (CA)
>
> Peak Construction Ltd was main contractor on a contract to construct a multi-storey block of flats for Liverpool Corporation. As a result of defective work by nominated sub-contractor McKinney Foundations, work on the main contract was halted for 58 weeks, and the main contractor brought a claim against the sub-contractor for damages. The Official Referee, at first instance, found that the entire 58 weeks was delay caused by the nominated sub-contractor, and awarded £40,000 of damages, £10,000 of which was for rises in wage rates during the period. McKinney appealed, and the Court of Appeal found that the award of £10,000 could not be upheld as clause 27 of the main contract entitled Peak Construction to claim this from Liverpool Corporation right up until the time when the work was halted.

8 Payment

8.1 Under DB16, the contractor has the right to be paid sums properly due, and the employer is obliged to ensure that the contractor is so paid. There are several safeguards in place, introduced through the Housing Grants, Construction and Regeneration Act (HGCRA) 1996 as amended by the Local Democracy, Economic Development and Construction Act (LDEDCA) 1999, to ensure that the contractor is dealt with fairly, so it is sensible for the employer (in the absence of any independent certifier) to make every effort to interpret the provisions objectively.

8.2 The payment provisions in DB16 are quite complex. However, as many of the provisions were introduced to comply with the above Act, and are compulsory for all contracts to which the Act applies, the clauses should not be amended without taking expert advice.

8.3 In summary, two 'methods of payment' are set out in the contract particulars, and the employer will indicate which will be used prior to tendering the contract (if nothing is selected, Alternative B applies). The methods are 'by stages' (Alternative A) or 'periodically' (Alternative B). However, the terminology is at first glance a little misleading, as in both cases payment is made monthly. The difference between the two methods lies in the way the amount to be paid is calculated. Essentially, under Alternative A the contractor is only paid for stages that are complete at the time the valuation is made, whereas under Alternative B all work correctly carried out is valued. This is a significant change to the system under DB11, where under Alterative A the contractor was paid 'at completion of each stage' (which very likely would have resulted in longer intervals between payment).

8.4 The timing of payments depends on 'Interim Valuation Dates'. The first date is entered in the contract particulars, and subsequent dates are at monthly intervals (if no first date is entered, it is one month after the date of possession). If Alternative A is selected, details of the stages are set out either in the contract particulars or in a document identified in the contract particulars. In respect of design work for stage payments, the description of each stage should make allowance for design work to be completed by that stage. For periodic payments, provision is made for valuation of design work.

8.5 There is an optional provision for advance payment of the main contractor (cl 4.6). An entry must be made in the contract particulars to show whether or not the optional provision is to apply (the contract states that it is not applicable if the employer is a local or public authority). If advance payment is required, the amount will be entered as either a fixed sum or as a percentage of the contract sum. The entry must also show when payment is to be made to the contractor and when it is to be reimbursed to the employer. A bond may be required, and DB16 includes an advance payment bond in Part 1 of Schedule 6.

8.6 There is always a risk in making an advance payment on a construction contract, even when backed by a bond, and the procedure will inevitably cause extra expense to the employer. The employer should be quite clear as to what compensatory benefits, such as a reduction in the contract sum, will result before agreeing to any arrangement of this sort.

Timing of payments

8.7   The contractor is required to make applications for payment to the employer (an 'Interim Payment Application'), which must state the sum the contractor considers is due and the basis on which it was calculated (cl 4.7.3) and should be accompanied by such information as is specified in the employer's requirements (cl 4.7.4). If the contractor makes the application prior to or on the interim valuation date, the 'due date' is seven days after the valuation date. If the application arrives later than the interim valuation date, the due date is seven days after its receipt by the employer (cl 4.7.2 and 4.7.3). The employer must issue a payment notice within five days of the due date, stating the sum it considers due and the basis on which this was calculated (cl 4.7.5). The final date for payment is 14 days from the due date (cl 4.9.1). The overall effect of this system is that the contractor's interim payment application is a condition precedent to its right to be paid.

Ascertainment of amounts due

8.8   The interim payment is calculated as the total amount ascertained under clause 4.12 (for Alternative A) or 4.13 (for Alternative B). The amount ascertained under clause 4.12 or 4.13 can be summarised as follows.

8.9   A total of 97 per cent (or as entered in the contract particulars) of the following (i.e. items subject to retention):

- for Alternative A:
    - the cumulative value of stages completed (cl 4.12.1.1);
    - the value of changes or of other work referred to in clause 5.2 that is relevant to the interim payment (cl 4.12.1.2; includes changes, work to be treated as a change and provisional sum work; excludes amounts referred to in clause 4.12.2.4, i.e. certain reinstatement work);
    - where fluctuations Option C applies, the value of certain site materials (cl 4.12.1.4);
- for Alternative B:
    - total value of work properly executed, including any design work (cl 4.13.1.1);
    - the value of changes or of other work properly executed and referred to in clause 5.2 (cl 4.13.1.1; includes changes, work to be treated as a change and provisional sum work);
    - total value of site materials (i.e. materials and goods properly on site) (cl 4.13.1.2);
- plus, for A and B:
    - value of 'listed items' (cl 4.12.1.3/4.13.1.3).

8.10 The above amounts are, for both alternatives, to be adjusted in accordance with any fluctuations provisions and with any acceleration quotation that has been agreed (cl 4.12.1/4.13.1).

8.11 A total of 100 per cent of the following items, if applicable:

- for Alternative A and Alternative B (although set out separately in the form, these appear to be identical for both alternatives):

- costs associated with clause 2.5.2 (additional insurance premiums), clause 2.20 (royalties), clause 3.12 (opening up and tests), clauses 6.10.2, 6.10.3 and 6.11.3 (terrorism cover), clause 6.12.2 (employer's insurance default) or clause 6.20 (joint fire code; cl 4.12.2.1/4.13.2.1);
- amounts payable under clause 4.11.2 (costs and expense reasonably incurred following suspension; cl 4.12.2.2/4.13.2.2);
- loss and/or expense due under clause 4.19.1 (cl 4.12.2.3/4.13.2.3);
- sums for restoration of damaged work under Insurance Option B or C, or under Option A to the extent that it is to be treated as a change (cl 4.12.2.4/4.13.2.4);
- amounts payable under any applicable fluctuations provision that have not been adjusted under clause 4.12.1/4.13.1 (cl 4.12.2.5/4.13.2.5).

8.12 Before reaching the total gross valuation, the contract requires some deductions to be made if applicable. These can be summarised as (for both Alternatives A and B):

- any amounts deductable under clause 2.35 (deductions in respect of defects agreed not to be made good), clause 3.6 (costs incurred by the employer where instructions not complied with), clause 6.12.2 (contractor's insurance default) or clause 6.19.2 (joint fire code) (cl 4.12.3.1/4.13.3.1);
- amounts allowable by the contractor in respect of clause 6.10.2 (terrorism cover) or any applicable fluctuations (cl 4.12.3.2/4.13.3.2).

Value of work

8.13 Under Alternative A, the value of work will be the total of the amounts set out in the contract particulars for all completed stages, with no amounts included for any partially completed stages. There will be no need to revalue the work, as it will be irrelevant whether or not the apportionment of the contract sum allocated to each stage actually correlates with any contract sum analysis that the contractor has provided at tender. Under Alternative B, there is no requirement that the value of the work properly executed, etc. is to be calculated using the rates shown in the contractor's contract sum analysis. However, it would be usual to base the application on this analysis, and it would be open to the employer to disagree with the application if it was made on some quite different basis.

8.14 In both Alternatives A and B, the employer should be careful not to pay for any work that appears not to have been properly executed, whether or not it is about to be remedied or the retention is adequate to cover remedial work. Clause 1.9 makes it clear that no interim payment is conclusive evidence that the work and materials paid for are in accordance with the contract; therefore, if work has been included in a payment and subsequently proves to be defective, the value can be omitted from the next payment. However, it is unlikely that this can be done to the extent that it produces a negative amount, i.e. a repayment by the contractor. Clause 4.7.1 refers to payments by the employer to the contractor and, unlike the final payment provisions, does not refer to a reverse payment. In addition, clause 4.10.3 states that the amount 'may be zero' but does not refer to negative amounts. This clause derives from the LDEDCA 2009 and, on balance, it seems unlikely that the contract will be interpreted to allow for negative payments. If such a situation arises, it may be sensible for the employer to seek legal advice, particularly if the amount is significant.

8.15   With respect to Alternative A, it is notable that there is no express proviso that the work in the relevant stage should have been properly completed, but it is suggested that the employer would be entitled to withhold payment for a stage where there were patent defects in the works (i.e. a court would construe the provision this way). If work which has been included in a payment subsequently proves to be defective, the employer may have to exercise its common law right of abatement to make a deduction from the payment due, provided it has issued the necessary notices.

Unfixed materials

8.16   Under Alternative B (but not Alternative A, except for specific items where Fluctuations Option C applies, cl 4.12.1.4), the payment should include materials which have been delivered to the site but not yet incorporated in the works (cl 4.13.1.2). In spite of detailed provisions aimed at protecting the employer, there remains some risk in including these items. Under common law, once materials have been built in, they will normally become the property of the owner of the land, irrespective of whether or not they have been paid for by the contractor. This would be the case even if there were a retention of title clause in the contract with the sub-contractor or supplier. A retention of title clause is one which stipulates that the goods sold do not become the property of the purchaser until they have been paid for, even if they are in the possession of the purchaser.

8.17   The employer could be at risk, however, where materials have not yet been built in, even where the materials have been certified and paid for. The contractor might not actually own the materials paid for because of a retention of title clause in the sale of materials contract. Under the Sale of Goods Act 1979, sections 16 to 19, property in goods normally passes when the purchaser takes possession of them, but a retention of title clause will be effective between a supplier and a contractor even where the contractor has been paid for the goods, provided they have not yet been built in. It should be noted, however, that the employer may have some protection through the operation of section 25 of the Act, which in some circumstances allows the employer to treat the contractor as having authority to transfer the title in the goods, even though this may not in fact be the case (*Archivent* v *Strathclyde Regional Council*).

> *Archivent Sales & Developments Ltd* v *Strathclyde Regional Council* (1984) 27 BLR 98
> (Court of Session, Outer House)
>
> Archivent agreed to sell a number of ventilators to a contractor who was building a primary school for Strathclyde Regional Council. The contract of sale included the term 'Until payment of the price in full is received by the company, the property in the goods supplied by the company shall not pass to the customer'. The ventilators were delivered and included in a certificate issued under the main contract (JCT63), which was paid. The contractor went into receivership before paying Archivent, which claimed against the Council for the return of the ventilators or a sum representing their value. The Council claimed that section 25(1) of the Sale of Goods Act 1979 operated to give it unimpeachable title. The judge found for the local authority. Even though the clause in the sub-contract successfully retained the title for the sub-contractor, the employer was entitled to the benefit of section 25(1) of the Sale of Goods Act 1979. The contractor was in possession of the ventilators and had ostensible authority to pass the title on to the employer, who had purchased them in good faith.

8.18   Another risk relating to rightful ownership occurs where the contractor fails to pay a domestic sub-contractor who has purchased materials, and the sub-contractor claims ownership of the unfixed materials. Here the risk may be higher, as a work and materials contract is not governed by the Sale of Goods Act 1979. Therefore, there can be no assumption that property in the materials would pass on possession.

8.19   DB16 attempts to deal with the issues surrounding ownership in several ways. First, unfixed materials and goods, which have been delivered to the site and are intended for the works, may not be removed without the written consent of the employer (cl 2.21). Removal would be a breach of contract and therefore the employer could claim from the contractor for any losses suffered through unauthorised removal. This would apply even though the materials or goods may not yet have been included in any payment. Second, unfixed materials and goods, either on or off site, which have been included in an interim payment will become the property of the employer, and the contractor is thereby prevented from disputing ownership (cl 2.21 and 2.22).

8.20   Clause 2.21 of the main contract, however, is binding only between the parties to the contract, and does not place obligations on any sub-contractor. The employer faces the risk that, if the contractor becomes insolvent, a sub-contractor or supplier may still have a rightful claim to ownership of the unfixed goods, even though they have been paid for by the employer (see *Dawber Williamson Roofing* v *Humberside County Council* below). The main contract therefore requires that all sub-contracts include similar clauses to clause 2.21 regarding non-removal from site, and ownership passing upon payment (cl 3.4.2.1). Sub-contracts must also include a clause stating that, once materials and goods have been certified and paid for under the main contract, they become the property of the employer and that the sub-contractor 'shall not deny' this (cl 3.4.2.1.1). This clause would operate even where the main contractor has become insolvent. Even this provision might not protect the employer in some circumstances because, if the sub-contractor does not have 'good title', it cannot pass it on. Thus, for example, it might not prevent a sub-subcontractor claiming ownership.

> *Dawber Williamson Roofing Ltd* v *Humberside County Council* (1979) 14 BLR 70
>
> The plaintiff entered into a sub-contract with Taylor and Coulbeck Ltd (T&C) to supply and fix roofing slates. The main contractor's contract with the defendant was on JCT63. By clause 1 of its sub-contract (which was on DOM/1), the plaintiff was deemed to have notice of all the provisions of the main contract, but it contained no other provisions as to when property was to pass. The plaintiff delivered 16 tons of roofing slates to the site, which were included in an interim certificate, which was paid by the defendant. T&C then went into liquidation without paying the sub-contractor, which brought a claim for the amount or, alternatively, the return of the slates. The judge allowed the claim, holding that clause 14 of JCT63 could only transfer property where the main contractor had a good title. (The difference between this and the *Archivent* case cited above is that in this case the sub-contract was a contract for work and materials, to which the Sale of Goods Act 1979 did not apply.) Provisions within clause 3.4 of DB16 now deal with the problem illustrated by this case.

8.21   Under the DB16 Alternative B provisions, the employer is obliged to include the unfixed materials payment, even though a limited risk remains, provided that the materials are not prematurely delivered to site and are adequately protected. Employers should pay careful attention to the exact wording of this condition.

'Listed items'

8.22   Alternatives A and B both refer to provisions for 'listed items'. These provisions allow for the contractor to be paid for materials or goods prior to their delivery to site (cl 4.15). If this is to apply, then a list of these items must be attached to the employer's requirements. The listed items may or may not be 'uniquely identified' (i.e. materials or goods or items prefabricated for inclusion in the works). The value of items listed must be included in an interim payment prior to delivery on site, provided that certain preconditions are fulfilled:

- the listed items are in accordance with the contract (cl 4.15.1);
- the contractor has provided reasonable proof that the property is vested in it (cl 4.15.2.1);
- the contractor provides proof that the items are insured against specified perils until delivery on site (cl 4.15.2.2).
- the listed items are 'set apart' or clearly marked and identified (cl 4.15.3);
- if the item is not 'uniquely identified' or if required in the contract particulars, the contractor has provided a bond (cl 4.15.4, 4.15.5).

8.23   The employer has no obligation to pay for any off-site items, other than those that have been listed. The employer should, therefore, be careful to avoid paying for any unlisted off-site materials, as only listed items are covered by the clause 4.15 safeguards regarding ownership and responsibility for loss or damage.

Costs and expenses due to suspension

8.24   Clause 4.11.2 states that 'Where the Contractor exercises his right of suspension under clause 4.11.1, he shall be entitled to a reasonable amount in respect of costs and expenses reasonably incurred by him as a result of exercising the right'. These amounts are also to be included in interim valuations. The phrase 'costs and expenses' is taken from the LDEDCA 2009, and suggests something more limited than the range of losses that could be claimed under a clause 4.19 loss and/or expense claim; for example, that it is limited to direct and ascertainable costs. However, a valid suspension would be due to a breach of contract by the employer, for which the contractor would be able to claim damages at common law, so it may be sensible for administrators not to interpret this clause too strictly.

Other items in the gross valuation

8.25   As noted above, in addition to the value of work properly executed and of materials properly on the site, clauses 4.12 and 4.13 list other amounts that must be included in the gross valuation for an interim payment. For example, payments ascertained as due to the contractor under clause 4.19 (direct loss and/or expense) are to be added to the contract sum (cl 4.12.2.3/ cl 4.13.2.3). Where adjustments to the contract sum have been made in accordance with the terms of the contract then these must be taken into account at the next interim payment (cl 4.3).

Deductions from the gross valuation

Withheld percentage

8.26 Some of the items that must be included in the gross valuation are subject to a reduction of 3 per cent, or any other amount entered in the contract particulars (cl 4.18). As an alternative, the contractor may be required to provide a retention bond (cl 4.17). If this is to be used, the contract particulars must state that clause 4.17 is to apply, and the details must be set out at tender stage. DB16 includes a form of bond in Part 3 of Schedule 6.

8.27 Where a retention bond is not used, the retention should be applied as appropriate to interim valuations. The rate is 3 per cent (unless another figure is inserted in the contract particulars) for all payments up till practical completion (cl 4.18.1). Half of the retention rate is withheld from certificates during the rectification period (cl 4.18.2; effectively half of the withheld amount is released in the first payment after practical completion). The remaining retention is released with the final payment. The employer is trustee of the withheld percentage for the contractor (cl 4.16.1), and must keep it in a separate bank account (unless the employer is a local or public authority), but is not obliged to invest it for the contractor, and will have the benefit of any interest which accrues (cl 4.16.2).

Advance payments and bond

8.28 If the parties have elected to use the advance payment provisions, the payment indicated in the contract particulars is made to the contractor before the first interim payment is due, but only after the contractor has provided the bond required (cl 4.6; DB16 includes a form of bond in Part 1 of Schedule 6). Details of when the reimbursements are to take place will also be set out in the contract particulars and could, for example, occur in stages throughout the project. The reimbursement is deducted from the gross valuation under the relevant payment. It should be noted that the amount to be deducted each month should be a cumulative total of the reimbursements, with the final deduction equalling the original advance payment, and that final deduction being made from all subsequent valuations.

Payment procedure

8.29 As noted in paragraph 8.7, the final date for payment is 14 days from 'its due date' (cl 4.9.1). The amount to be paid on that date is subject to provisions regarding notices, which are required by the Housing Grants, Construction and Regeneration Act 1996 (as amended) and which apply also to the final payment. Under clause 4.7.5, no later than five days after the due date for payment the employer must give the contractor a notice (the 'Payment Notice') which states how much it intends to pay. The notice is required whether or not the employer intends to make any deduction from the amount applied for (see Figure 8.1). If the employer fails to issue a payment notice, clause 4.9.3 states that 'the Employer shall, subject to any Pay Less Notice under clause 4.9.5, pay the Contractor the sum stated as due in the Interim Payment Application'.

8.30 Should the employer decide that it wishes to withhold any amount from the sum stated in its payment notice (or, if no payment notice was issued, from the amount in the interim payment application), then it must give written notice of this intention no later than five

Figure 8.1 Interim payment procedure

- Valuation date (monthly, entered in contract particulars)
- Due date, 7 days after later of:
  • valuation date (4.7.2)
  • date of receipt by employer of contractor's application (4.7.3)
- Final date for payment
- 14 days
- 5 days
- 4.7.3 Contractor **shall** make an interim application
- 4.7.5 Employer **shall** issue a payment notice to the contractor
- 4.9.5 Employer may issue a pay less notice in accordance with 4.10.1
- 5 days
- 4.9.3 If payment notice not issued in accordance with 4.9.2, amount due to be the sum stated in the interim application

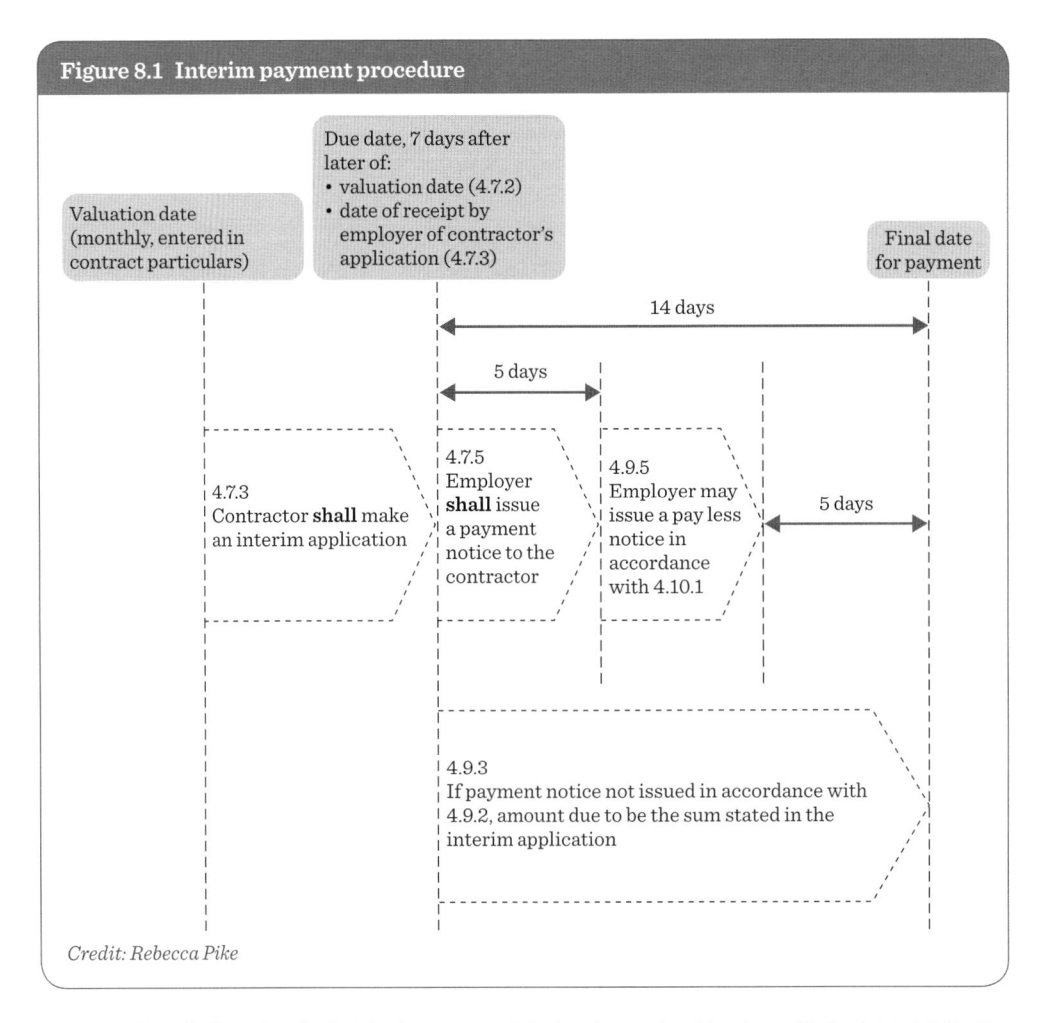

Credit: Rebecca Pike

days before the final date for payment, in the form of a 'Pay Less Notice' (cl 4.9.5). The pay less notice should state the sum considered to be due and the basis on which that sum has been calculated (cl 4.10.1). It is suggested that if it is known in advance that a deduction is intended from the amount applied for, then the payment notice should take account of this, and the pay less notice will then be unnecessary. Contractors should note that, always provided the payment notice includes all the necessary information required under clause 4.7.5, there is no requirement for the employer to issue a pay less notice simply because it disagrees with the amount shown in the contractor's application.

Deductions

8.31　The contract expressly gives the employer the right to make certain deductions from any amount due to the contractor. These are listed in clauses 4.12.3 and 4.13.3; for example deductions from the contract sum where defective work has been accepted (see paragraphs 6.41 and 6.60). These should normally be deducted before reaching the gross valuation in an application for payment or payment notice. It is suggested, however, that the pay less notice may also make these deductions, if not already accounted for in the payment notice. For example, if not all information was available at the time of the payment notice, or new circumstances develop following the payment notice, then the deduction could be effected by means of a pay less notice. The only other deduction authorised by the contract, and not set out in clauses 4.12.3 and 4.13.3, arises in respect of payment or allowance of liquidated damages (cl 2.29.2.2). Should the employer wish to use the pay less notice to make any other claim, for example to set off an amount it believes it is owed from another project, it should not do so without taking legal advice.

8.32   It should be noted that no application for payment, and no payment by the employer, offers conclusive proof of the amount due, except for the final statement, as discussed below (cl 1.9). Despite having paid a specific amount, it would always subsequently be open to the employer to disagree with this figure. The employer may decide to make a deduction from a further payment, or to raise the matter in adjudication. In particular, having made payment does not preclude the employer from exercising any common law right, such as abatement.

Employer's obligation to pay

8.33   As stated above, the employer is required to pay the amount due by the final payment date. The amount paid should be 'not less than' the amount set out in any employer's pay less notice (clause 4.9.5). If no pay less notice has been given, it will be the amount stated in the employer's payment notice or, if none has been issued, the amount in the contractor's application.

8.34   In addition to the contractual rights to make deductions by means of a notice as discussed above, the employer may have other rights to withhold payment under common law. Prior to the HGCRA 1996, it was clear that, if the employer had an arguable case that the certificate included work which was defective, and therefore had been overvalued, then, rather than paying the full amount, the employer could have raised the losses due to the defects either as a counterclaim in any action brought by the contractor or as a defence to the claim. The latter process is often termed 'abatement' by lawyers.

8.35   It is now generally agreed that, in cases where such a right may exist, it can only be exercised through the use of the 'Pay Less Notice' procedure, as discussed above. The employer would therefore be unable to withhold amounts to cover any defective work included in a payment notice unless the deduction is covered by a notice (*Rupert Morgan Building Services (LLC) Ltd* v *David Jervis and Harriet Jervis*). The rights of the employer when defects appear after the expiry of the time limits for notices but before the final date for payment are unclear, but it is arguable that, in such situations, the employer would retain a right to abatement of the amount due.

> *Rupert Morgan Building Services (LLC) Ltd* v *David Jervis and Harriet Jervis* [2004] BLR 18 (CA)
>
> A couple engaged a builder to carry out work on their cottage, under a contract on the standard form published by the Architecture and Surveying Institute (ASI). The seventh interim certificate was for a sum of around £44,000 plus VAT. The clients accepted that part of that amount was payable but disputed the balance amounting to some £27,000. The builder sought

> summary judgment for the balance. The clients did not give 'a notice of intention to withhold payment' before 'the prescribed period before the final date for payment'. The builder contended that it followed, by virtue of HGCRA 1996 section 111(1), that the clients 'may not withhold payment'. The clients maintained that it was open to them, by way of defence, to prove that the items of work which made up the unpaid balance were either not done at all, or were duplications of items already paid or were charged as extras when they were within the original contract, or represented 'snagging' for works already done and paid for. The Court of Appeal determined that, in the absence of an effective withholding notice, the employer has no right of set-off against a contract administrator's certificate.

Contractor's position if the amount applied for is not paid

8.36 DB16 includes several provisions which protect the contractor if the employer fails to pay the contractor any amounts due. Clause 4.9.6 makes provision for simple interest on late payments. This is set at 5 per cent over the base rate of the Bank of England, and the interest accrues from the final date for payment until the amount is paid. Similar provisions are included for the final payment. (It should be noted that if the provision were deleted the contractor would normally have a statutory right to interest under the Late Payment of Commercial Debts (Interest) Act 1998.) If the employer makes a valid deduction following a notice, it is suggested that interest would not be due on this amount. The contract makes it clear that acceptance of a payment of interest does not constitute a waiver by the contractor of its right to proper payment of the amount due (cl 4.9.7).

8.37 The contractor is also given a 'right of suspension' under clause 4.11. This right is required by the HGCRA 1996. If the employer fails to pay the contractor by the final date for payment, then the contractor has a right to suspend performance of all its obligations under the contract, which would include not only the carrying out of the work but, for example, could also extend to any insurance obligations. This sum payable is to be in accordance with cl 4.9, therefore the contractor may not suspend work if a pay less notice has been given by the employer and the employer has paid the amount set out in that notice. In order to exercise the right of suspension, the contractor must have given the employer written notice of its intention and stated the grounds for the suspension, and the employer's default must have continued for a further seven days (cl 4.11.1). The contractor must resume work when the payment is made. Under these circumstances, the suspension would not give the employer the right to terminate the contractor's employment. Any delay caused by the suspension could be a relevant event (cl 2.26.5) and the contractor 'shall be entitled to a reasonable amount in respect of costs and expenses reasonably incurred by him as a result of exercising the right' (cl 4.11.2).

8.38 The contractor also has the right to terminate the contract if the employer does not pay any amounts due (cl 8.9.1.1). The contractor must give notice of this intention, and must specify the default, as required by the contract (see paragraph 10.26).

Interim payment after practical completion

8.39 Under both Alternatives A and B, an interim payment will be made at the end of the period in which practical completion falls. As noted above, the amount due will include 98.5 per cent of the value referred to in clause 4.12.1/4.13.1. The contract provides for further payments at monthly intervals between practical completion and the final payment (cl 4.7.2). In some cases no amounts may be due, but a payment certificate should still be issued, even if the amount is zero (cl 4.10.3). A payment may arise from, for example, a claim for loss and/or expense or the amount due as a result of the valuation of a change that had not been resolved prior to practical completion.

Final payment

8.40 Following practical completion the contractor is required to send a final statement to the employer for agreement, together with such supporting information as the employer may reasonably require (cl 4.24.1; see Figure 8.2). The final statement sets out the adjustments to the contract sum already made, the contract sum as so adjusted, the sum of amounts already paid, the balance resulting from the two and 'the basis on which that amount has been calculated' (cl 4.24.2). The adjustments will include amounts relating to:

- provisional sums;
- corrections of divergences;
- changes in statutory requirements;
- employer's instructions effecting a change;
- insurance;
- loss and/or expense;
- fluctuations;
- costs and expense reasonably incurred under clause 4.11.2 following suspension.

8.41 If the contractor fails to provide the final statement within three months of practical completion, the employer may take steps to prepare this document (cl 4.24.3). The employer must issue a two-month notice to the contractor stating its intention to do so and, once that period has elapsed, may issue the employer's final statement at any time, assuming that in the meantime the contractor has not submitted a final statement (cl 4.24.3 and 4.24.4).

8.42 The final statement can be for a negative amount – in other words, it can state that payment is due from the contractor to the employer. This may occur, for example, where there are revisions to the calculation of adjustments, or where an agreement has been reached regarding the acceptance of defective work.

8.43 The final date for payment of the balance is 14 days from the due date (clause 4.9.1). The due date is one month after either the date of submission of the final statement, the date of the end of the defects period, or the date stated in the notice of completion of making good, whichever of these occurs last (cl 4.24.5).

Conclusive effect of final statement

8.44 The final statement becomes conclusive as to the balance due for the final payment (cl 4.24.6 and 1.8.1). (More precisely, it acquires the status of 'conclusive evidence' of the amount due in that, although a party could later initiate a claim regarding the amount due, it would not be entitled to bring any evidence to challenge the amount in the statement.)

Figure 8.2 Final payment procedure

Practical completion → 3 months →

4.24.5 Whichever occurs last:
1. end of rectification period
2. date stated in the notice of completion of making good
3. date of submission of final statement

4.24.1 Contractor **shall** submit the final statement

→ 1 month → **Due date** → 28 days → **Final date for payment**

4.9.5 Either party **may** issue a pay less notice in accordance with 4.10.1 — [A] 5 days / 5 days

→ 2 months →

4.24.3 If the contractor fails to submit a final statement, the employer may give notice to the contractor that the employer will issue a final statement in lieu of that of the contractor if not supplied within 2 months

→ 1 month → **Due date** → 28 days → **Adjusted final date for payment**

4.9.5 Either party **may** issue a pay less notice in accordance with 4.10.1 — [A] 5 days / 5 days

[B]

[A] 4.8 Payer **shall** issue a payment notice to the payee.
[B] 4.24.4 If the contractor does not exercise its rights to submit a final statement, the employer may at any time issue an employer's final statement to the contractor.

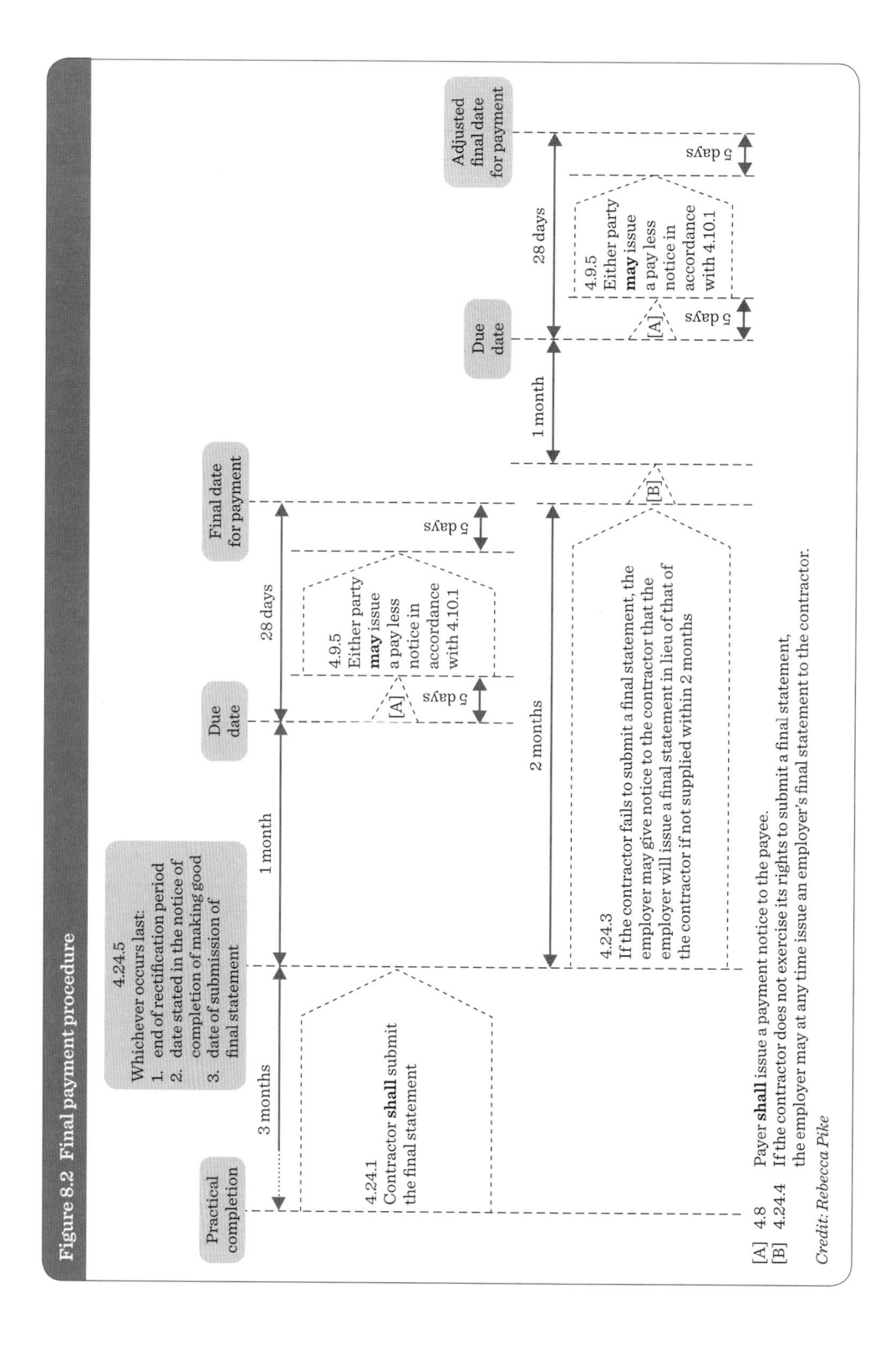

Credit: Rebecca Pike

The conclusive status of the final statement starts on the due date, but is said to be 'except to the extent that' the employer (in the case of a statement prepared by the contractor) or the contractor (in the case of a statement prepared by the employer) issues a notice disputing anything in the relevant statement prior to the due date (cl 4.24.6). This exception is 'subject to clause 1.8.2' which states that the conclusiveness of the statement 'shall in relation to the subject matter of any adjudication, arbitration, or other proceedings be suspended pending the conclusion of such proceedings ... where those proceedings are commenced before or within 28 days after the date of issue of the relevant statement'. It therefore appears that two notices are required in order to prevent the statement from becoming conclusive, an initial one prior to the due date, and a formal initiation of dispute resolution proceedings within 28 days of the date of issue of the final statement.

8.45   It should be noted that the suspension is only 'to the extent that' the statement is disputed. The final statement, when it becomes conclusive of amounts due between the parties, is also conclusive with respect to the extensions of time awarded and loss and/or expense ascertained, and the contractor is prevented from seeking to raise any further claims regarding these matters (cl 1.8.1.2 and 1.8.1.3). It is also conclusive evidence, where matters have been expressly stated to be for the approval of the employer, that they have been approved, but, those matters aside, it is not conclusive evidence that any other materials, workmanship, etc. comply with any contractual requirement (cl 1.8.1.2, see also the case of *London Borough of Barking & Dagenham* v *Terrapin Construction Ltd* discussed at paragraph 4.15). It would also not be conclusive evidence that the design complies with the contractual requirements. If only one of these items is disputed, the statement will become conclusive as to the others.

8.46   The suspension will apply to adjudication provisions, if they are commenced within the stipulated time limits. The decision of an adjudicator is binding on the parties, unless and until it is challenged by means of arbitration or litigation. In order to achieve some finality, DB16 states that the suspension on conclusiveness will cease unless such proceedings are commenced within 28 days of the adjudicator's decision (cl 1.8.2.2). To deal with the common situation where a party initiates proceedings, but then takes no further action, the contract sets a long stop of 12 months from the issue of the relevant statement – if no further action is taken the statement then becomes conclusive (cl 1.8.2.3).

9 Indemnity and insurance

9.1 One of the most important functions of a building contract is clearly to allocate liability for the risks inherent in any construction operation, i.e. the risks of accident, injury and damage to property. Should any such incidents occur, it is vital that there should be no room for dispute about who is liable for the losses, and that all concerned should be clear about what procedural steps must be taken. Ambiguity in the contract can only lead to confusion and delays, which will benefit neither party.

9.2 Normally, a building contract will set out the specific events for which the contractor is liable, and require the contractor to indemnify the employer in respect of the resultant losses, for example for injury to persons or damage to neighbouring property. In DB16 these liabilities are allocated under clauses 6.1 and 6.2. Clause 6.1 makes the contractor liable for, and requires indemnification of the employer against, losses or claims due to injury to or death of persons, or damage to neighbouring property which has been caused by the contractor's negligence. The indemnity protects the employer in that, if an injured party brings an action against the employer, rather than against the contractor, the contractor will bear the consequences of the claim. In practice, the employer can either join the contractor as co-defendant or bring separate proceedings against the contractor.

9.3 In practice, the indemnities given to the employer by the contractor are quite worthless if the contractor has insufficient resources to meet the claims. DB16 therefore requires the contractor, under clause 6.4, to carry insurance cover to underwrite the indemnities required under clauses 6.1 and 6.2.

9.4 In addition to the requirement for insurance against claims arising in respect of persons and property, the contract contains alternative provisions for insurance of the works under clauses 6.7 to 6.11 and Schedule 3. There is also an optional provision requiring the contractor to take out insurance for non-negligent damage to property other than the works (cl 6.5).

Injury to persons and damage to property caused by the negligence of the contractor

9.5 Clause 6.4 requires the contractor to carry insurance to cover injury to persons and damage to property other than the works which arise from the carrying out of the works. The contractor must be able to provide evidence that this insurance has been taken out. If the contractor defaults, the employer may take out the insurance and recover the costs from the contractor as a debt (cl 6.4.1 and 6.12).

9.6 Clause 6.4.1.1 requires that the insurance in respect of personal injury or death of any person in a contract of service with the contractor should comply with 'all relevant legislation'. The contractor's liability in respect of personal injury or death of employees is met by an employer's liability policy. This has been compulsory since the Employer's Liability (Compulsory Insurance) Act 1969. The statutory requirement is for a cover level of £5 million, although in practice most standard policies provide cover of at least £10 million.

9.7 The contractor's liability in respect of third parties (death or personal injury and loss or damage to property including consequential loss) is met by its public liability policy. Insurers advocate insuring for a minimum of £2 million for any one occurrence, although a higher amount may be required by some clients. The contractor must insure the indemnities required under clauses 6.1 and 6.2 up to the amount stated in the contract particulars. However, liability at common law for claims by third parties is unlimited, and any amount specified in the contract is merely the employer's requirement in the interests of safeguarding against inadequacies and in no way limits the contractor's liability under clauses 6.1 and 6.2. Furthermore, it is recognised in footnote [47] to clause 6.4.1.2 that it may not always be possible to acquire insurance cover for all the indemnities required in clauses 6.1 and 6.2. For example, the insurance market has removed gradual pollution from its public liability policies. This in no way affects the contractor's duty to indemnify.

9.8 The liability and duty to indemnify are both subject to exceptions. In respect of liability for personal injury or death, these are qualified in that the contractor is not liable where injury or death is caused by an act of the employer or a person for whom the employer is responsible (cl 6.1).

9.9 In respect of damage to property, the contractor is only liable to the extent that the damage is caused by negligence or breach of statutory duty or other default of 'the Contractor or any Contractor's Person' (cl 6.2). The contractor is therefore liable only for losses caused by its own negligence. It is made clear in clause 6.3.4 that the definition of 'property' excludes the works, up to practical completion of the works or a section, except parts taken over by partial possession.

9.10 Clause 6.3.1 also excludes, where Insurance Option C applies, liability for 'any loss or damage to Existing Structures or to any of their contents required to be insured under that option that is caused by any of the risks or perils required or agreed to be insured against under that option'. This means that, where Insurance Option C is applicable, the contractor is not liable for losses insured under paragraph C.1 and caused by the listed perils, even where the damage is caused by the contractor's own negligence. This point is now expressly stated in clause 6.3.2 (where a paragraph C.1 replacement schedule is used, the liability is subject to exclusions and limitations set out under that schedule; cl 6.3.3). The clause 6.3.2 exclusion was inserted to clarify matters following a series of cases on older versions of the contract that reached the opposite conclusion (*National Trust v Haden Young*, *London Borough of Barking & Dagenham v Stamford Asphalt Co.*). It should be noted that the contractor might remain liable for some consequential losses (*Kruger Tissue v Frank Galliers*).

> *The National Trust for Places of Historic Interest and Natural Beauty v Haden Young Ltd*
> (1994) 72 BLR 1 (CA)
>
> The National Trust employed a contractor to carry out repair works to Uppark House, South Harting, West Sussex. The main contract was on terms substantially similar to MW80. Haden Young was sub-contractor for the renewal of lead work on the roof. During the course of the

Indemnity and insurance

works a fire broke out, causing extensive damage, which Haden Young admitted was caused by the negligence of its workforce, and the National Trust brought a claim for damages. Otton J found the sub-contractor liable at first instance, and that the employer's liability to insure under clause 5.4B only extended to matters not caused by negligence. Clauses 5.2 and 5.4B formed a coherent and mutually supportive structure. Haden Young appealed, but the appeal was dismissed. Although the Court of Appeal agreed that the sub-contractor was liable, it disagreed with the reasoning of the lower court, stating that there was no reason why there should not be an overlap, in other words why the employer should not be required to insure for matters for which the contractor was liable under clause 6.2. However, the damages recoverable from the contractor under clause 6.2 would be reduced by the amount recoverable by the employer under the clause 6.3B insurance.

London Borough of Barking & Dagenham v *Stamford Asphalt Co. Ltd* (1997) 82 BLR 25 (CA)

Barking & Dagenham employed a contractor to carry out repair works to a school. The main contract was on MW80, 1988 revision. Stamford was sub-contractor for the renewal of lead work on the roof. During the course of the works a fire broke out, causing extensive damage, which Stamford admitted was caused by the negligence of its workforce, and the Borough brought a claim for damages. The Court of Appeal found the contractor liable for the damage caused, preferring the reasoning of Otton J in *National Trust* v *Haden Young* to that of the Court of Appeal in that case. It should be noted that the wording of clause 6.2 (now cl 5.2) has been adjusted to make it clear that the contractor is not liable for damage to property insured under clause 6.3B (now cl 5.4B).

Kruger Tissue (Industrial) Ltd v *Frank Galliers Ltd* (1998) 57 Con LR 1

Damage was caused to the existing building and works by fire, assumed for the purposes of the case to be the result of the negligence of the contractor or sub-contractor. The construction work being carried out was on a JCT80 form. The employer brought a claim for loss of profits, increased cost of working and consultants' fees, all of which were consequential losses. Judge John Hicks decided that the employer's duty to insure for 'the full cost of reinstatement, repair or replacement' of the existing structure and the works under clause 22C (and therefore contractor's exemption from liability under clause 20.2) did not include such consequential losses. A claim could therefore be brought against the contractor for these losses. (Note that DB16 now provides for professional fees coverage to be required as part of the works insurance.)

Damage to property not caused by the negligence of the contractor

9.11 The liability for damage to adjoining buildings, where there has been no negligence on the part of the contractor, is not covered under clause 6.2. Subsidence or vibration resulting from the carrying out of the works might cause such damage even though the contractor has taken reasonable care. This risk may be quite high in certain projects on tight urban sites or in close proximity to old buildings. In such cases it may be advisable to take out a special policy for the benefit of the employer.

9.12 In DB16 there is an optional provision for this type of insurance under clause 6.5. If it is expected that the main contractor may be required to take out this insurance, then this must be stated in the employer's requirements, and the amount of cover entered in the contract particulars. The employer must then instruct the contractor to take out the policy, if it is required. The cost is added to the contract sum. The policy must be in joint names and placed with insurers approved by the employer. The policy and receipt are to be deposited with the employer.

9.13 This insurance is usually expensive, and subject to a great many exclusions. If it is required, then the policy must be effective at the start of the site operations, when demolition, excavation, etc. are carried out. The text of clause 6.5 was revised in 1996 to take account of the wording of model exclusions compiled by the Association of British Insurers. The policy should be checked by the employer's insurance advisers to ensure that any exclusions correlate with clause 6.5 and that the policy provides the cover that this clause requires.

Insurance of the works

9.14 There are three options for covering insurance of the works (Options A, B and C), and the applicable option should be selected in the contract particulars (cl 6.7). In all cases, the policies are to be in joint names, and cover must be maintained until practical completion of the works, or termination, if this should occur earlier. The 'Joint Names Policy' definition was reworded in 1996 to make it clear that, under the policy, the insurer does not have a right of subrogation to recover any of the monies from either of the named parties. The works insurance policies must, in addition, either cover sub-contractors or include a waiver of any rights of subrogation against them (cl 6.9.1). This coverage is in respect of specified perils only, and not the full range of risks covered by 'All Risks Insurance'.

9.15 Insurance Option A and Option B deal with the insurance of new building work and require 'All Risks' cover under joint names policies. A definition of 'All Risks' is given in clause 6.8 and refers to 'any physical loss or damage to work executed and Site Materials and against the reasonable cost of the removal and disposal of debris'. There is also a list of exclusions, which includes the cost necessary to repair, replace or rectify property which is defective, loss or damage due to defective design, loss or damage arising from war and hostilities and 'Excepted Risks' (except as provided by terrorism cover). Footnote [51] explains that cover should not be reduced beyond the exclusions set out in the definitions. It also points out that 'All Risks' cover that includes the risk of defective design, although not required, may be available. If the policy provided is likely to differ in any way from the requirements in the contract this must be discussed and agreed before the contract is entered into. Even in a so-called 'All Risks' insurance policy there may be further exclusions, and the employer's insurance advisers should carefully check the wording of each policy.

9.16 Option A insurance is taken out by the contractor and must be for the full reinstatement value of the works, including professional fees, to the extent entered in the contract particulars. The contractor is responsible for keeping the works fully covered and, in the event of underinsurance, will be liable for any shortfall in recovery from the insurers.

9.17 Option B insurance is taken out by the employer, and must be for the full reinstatement value of the works, including professional fees. The employer is responsible for keeping the works fully covered and, in the event of underinsurance, will be liable for any shortfall.

Indemnity and insurance 119

9.18   Option C is applicable where work is being carried out to existing buildings. It includes two insurances, both taken out by the employer. The existing structure and contents must be insured against 'Specified Perils' as defined in clause 6.8 (Option C.1). (The main difference between 'All Risks' and 'Specified Perils' is the omission in the latter definition of risks connected with impact, subsidence, theft or vandalism.) New works in, or extensions to, existing buildings must be covered by an 'All Risks' insurance policy (Option C.2) and, as with Options A and B, must be for the full reinstatement value of the works, including professional fees.

9.19   In cases where the employer may have difficulty in obtaining the joint names insurance for the existing building, which might be the case with tenants and homeowners, the contract now offers an option whereby the parties may 'disapply' Option C.1 and replace it with alternative provisions. This must be stated in the contract particulars, which must also describe the document in which the alternative provisions are set out. The employer will need to consider what these provisions might be before tenders are sought, and there are likely to be negotiations before the matter can be finalised. All relevant insurers should, of course, be consulted and (particularly for inexperienced employers) specialist advice may well be required.

Action following damage to the works

9.20   The procedure is similar under Insurance Options A, B and C. The contractor must notify the employer in writing of the details of the damage as soon as possible (cl 6.13.1). Although not required to do so by the contract, the contractor or employer, depending on which has taken out the policy, should also inform the insurers immediately on becoming aware of the damage. After any necessary inspection has been made by the insurers, the contractor is then obliged to make good the damage and continue with the works (cl 6.13.4). Under all three options, the contractor authorises the payment of all monies due under the insurance policy to be made directly to the employer (cl 6.13.3).

9.21   Clause 6.13.2 states that 'the occurrence of such loss or damage to executed work or Site Materials shall be disregarded in calculating any amounts payable to the Contractor'. Payments that have already been made are not affected by the occurrence of the damage. In addition, any work that was completed after the most recent payment, but was then subsequently damaged, should be included in the next payment.

9.22   Under Option A, the contractor must take out insurance for the full reinstatement value of the works, plus a percentage to cover professional fees if this is required in the contract particulars (A.1). The insurance monies paid to the employer, minus the part of it to cover professional fees, but without any removal of retention, should be included in separate reinstatement work statements as the work is carried out, issued at the same time as the usual payment notices (cl 6.13.5.1). If the amount paid by the insurers is less than it costs the contractor to rebuild the works, the contractor is not entitled to any additional payment (cl 6.13.5.3), except in limited circumstances related to a reduction in terrorism cover. The risk of any underinsurance therefore lies with the contractor.

9.23   Under Options B and C, the rebuilding, restoration or repair work is treated as a change (cl 6.13.6), and therefore the contractor is less at risk and the employer will have to bear any shortfall in the monies paid. Under clause 2.26.2.9 the contractor is entitled to an extension of time for delay caused by loss or damage due to any of the specified perils.

In addition, if the work is treated as an instruction under clause 3.9, the contractor may be entitled to an extension of time and to claim for loss and/or expense under clauses 2.26.1 and 4.21.1 (see paragraph 6.31). In all cases, the entitlement appears to extend even to cases where the damage was caused by the contractor's negligence.

9.24 Either party is given the right to terminate the employment of the contractor where there is extensive damage to existing structures, and if 'it is just and equitable' to do so, by means of a 28-day notice (cl 6.14). The question of termination might arise, for example, where an existing structure to which work is being carried out has been completely destroyed, and it would be unreasonable to expect the contractor to rebuild. If the other party disagrees and feels that the project should continue, it must invoke the dispute resolution procedures. If the contract is terminated, the provisions of clauses 8.12 (except for 8.12.3.5) will apply. It should be noted that this right is in addition to the right under clause 8.11 of either party to terminate the contractor's employment should the works be suspended for a period of two months due to loss or damage caused by any risk covered by the works insurance policy or by any excepted risk (cl 8.11.1.3).

Terrorism cover

9.25 Under clause 6.10 the contractor (where Insurance Option A applies) or the employer (where Insurance Options B or C applies) is required to take out terrorism cover. This can be done either as an extension to the joint names policy or as a separate joint names policy, and must be taken out in the same amount and for the required period of the joint names policy. 'Terrorism Cover' is defined as 'Pool Re Cover or other insurance against loss or damage to work executed and Site Materials (and/or, for the purposes of clause 6.11.1, to an Existing Structure and/or its contents) caused by or resulting from terrorism' (cl 6.8). Pool Re Cover is also a defined term, and stands for 'such insurance against loss or damage to work executed and Site Materials caused by or resulting from terrorism as is from time to time generally available from insurers who are members of the Pool Reinsurance Company Limited scheme or of any similar successor scheme'. Although there have been difficulties in the past in obtaining such cover, at the time of writing the insurance market is prepared to cover terrorism risks. If cover is not available, or is likely to differ from the contractual requirements, then, as with insurance generally, the details must be agreed before the contract is signed.

9.26 If the insurers named in the joint names policy decide to withdraw this cover and notify either party, that party must immediately notify the other that terrorism cover has ceased (cl 6.11.1). The employer must then decide whether or not it wishes to continue with the works, and notify the contractor accordingly (cl 6.11.2). If the employer decides to terminate the contractor's employment, the provisions of clause 8.12 apply (cl 6.11.4, see Chapter 10). Otherwise, should any damage be caused by terrorism, this is to be dealt with under clause 6.13 or 6.14 as appropriate (cl 6.11.5).

Professional indemnity insurance

9.27 Under DB16 the contractor is required to carry professional indemnity insurance (cl 6.15). The level and amount of cover must be inserted in the contract particulars. If no level is inserted, it will be 'the aggregate amount for any one period of insurance', and if no amount is stated then no insurance will be required. There is a provision for inserting a level of cover for pollution or contamination claims. In addition, if the expiry period is to be 12 years from practical completion then this must be indicated, otherwise the period will be 6 years. The insurance must be taken out immediately following the execution of the contract, and maintained until the end of the stipulated expiry period. The contractor must provide evidence of the insurance if required (cl 6.15.3).

Joint Fire Code

9.28 The Joint Fire Code (cl 6.17 to 6.20) is designed to reduce the incidence of fire on construction sites. It is an optional provision selected in the contract particulars but, as compliance with the code may reduce the cost of some insurance policies, its inclusion should be carefully considered. If it is included, then both parties undertake to comply with the code and to ensure that those employed by them also comply.

9.29 If a breach of the Joint Fire Code occurs, the insurers may give notice to either the employer or the contractor of remedial measures they require and the dates by which these must be put into effect. If either party receives such a notice, they must copy it to the other (cl 6.19.1). If the notice sets out measures which conform to the contractor's existing obligations under the contract, then the contractor should put the measures in place (cl 6.19.1). If the contractor does not comply with the notice within seven days, the employer may employ and pay others to effect such compliance (cl 6.19.2).

Other insurance

9.30 There remain risks to the employer that are not covered by the DB16 insurance provisions. For example, if the contractor is caused delay by one of the specified perils, an extension of time would normally be awarded under clause 2.26.2.9 and the employer will not be able to claim liquidated damages from the contractor for that period. There will therefore be a loss to the employer. Should the employer wish to be insured against this loss of liquidated damages, then special provisions must be made, as there is nothing in DB16 which deals with such loss. The possible risks should be explained to the employer, but it should be noted that there are often problems with such insurance, as liquidated damages are payable without proof and, traditionally, insurers only pay on proof of actual loss. As a result, only one or two firms are currently willing to offer cover, and the price tends to be high.

9.31 There are other forms of insurance which are not covered by the provisions of DB16, and which the employer might wish to consider. The employer is the party best placed to assess possible loss. Where there are likely to be business or other economic losses, then these can be covered, albeit at a price. It is also possible to insure against defects occurring in the buildings by means of project-related insurance. This insurance is relatively expensive and limited to a ten-year 'decennial' loss. Irrespective of blame, it means that funds are available to remedy the defects, which will occur most often in the first eight years of the life of a building. Project-related insurance should include subrogation waiver, and does not reduce the need for professional indemnity cover.

10 Default and termination

10.1   In design and build procurement, the consequences of any serious breakdown of relations resulting from default are likely to be even more complex and difficult to resolve than under traditional procurement. Should the contractor's employment cease, the employer will be faced with a half-erected structure to complete, almost certainly with an incomplete design, and without the continuing services of the person responsible for completing it. If the contractor is unable to complete the project, it will remain liable for its design, even where the construction is completed by others. Clearly, this situation is very unsatisfactory for all parties and should be avoided if at all possible. However, problems can occur which are so serious that the other party may prefer not to continue with the contract, and for these more serious situations the contract contains provisions allowing for termination of the employment of the contractor.

Repudiation or termination

10.2   In any contract, where the behaviour of one party makes it difficult or impossible for the other to carry out its contractual obligations, the injured party might allege prevention of performance and sue, either for damages or a *quantum meruit*. This could occur in construction, for example, where the employer refuses to allow the contractor access to part of the site.

10.3   Where it is impossible to expect further performance from a party, then the injured party may claim that the contract has been repudiated. Repudiation occurs when one party makes it clear that they no longer intend to be bound by the provisions of the contract. This intention might be expressly stated, or implied by the party's behaviour.

10.4   Most JCT contracts include termination clauses, which provide for the effective termination of the employment of the contractor in circumstances which may amount to, or which may fall short of, repudiation. The provisions in DB16 are very similar to those in SBC16. It should be noted that such termination is only of the contractor's employment under the contract, and is not termination of the contract itself. The parties remain bound by the contract, and can bring actions for losses suffered through breach of its terms.

10.5   If repudiation occurs, it is unnecessary to invoke a termination clause, since the injured party can accept the repudiation and bring the contract to an end. However, the termination provisions are useful in setting out the exact circumstances, procedures and consequences of the termination of employment. These procedures must be followed with great caution because, if they are not administered strictly in accordance with the terms of the contract, this in itself could amount to a repudiation. This, in turn, might give the other party the right to treat the contract as at an end and claim damages.

10.6   Termination can be initiated by the employer (cl 8.4) in the event of specified defaults by the contractor, such as suspending the works or failing to comply with the CDM Regulations, or in the event of the insolvency of the contractor. Termination can be initiated by the contractor (cl 8.9) in the event of specified defaults by the employer, such as failure to pay an amount due, or where specified events result in the suspension of work beyond a period to be entered in the contract particulars. Termination might also follow the insolvency of the employer. In the event of neutral causes, which bring about the suspension of the uncompleted works for the period listed in the contract particulars, the right of termination can be exercised by either party (cl 8.11).

Termination by the employer

10.7 The contract provides for termination of the contractor's employment under certain circumstances. DB16 expressly states that the right to terminate the contractor's employment is 'without prejudice to any other rights and remedies' (cl 8.3.1). This

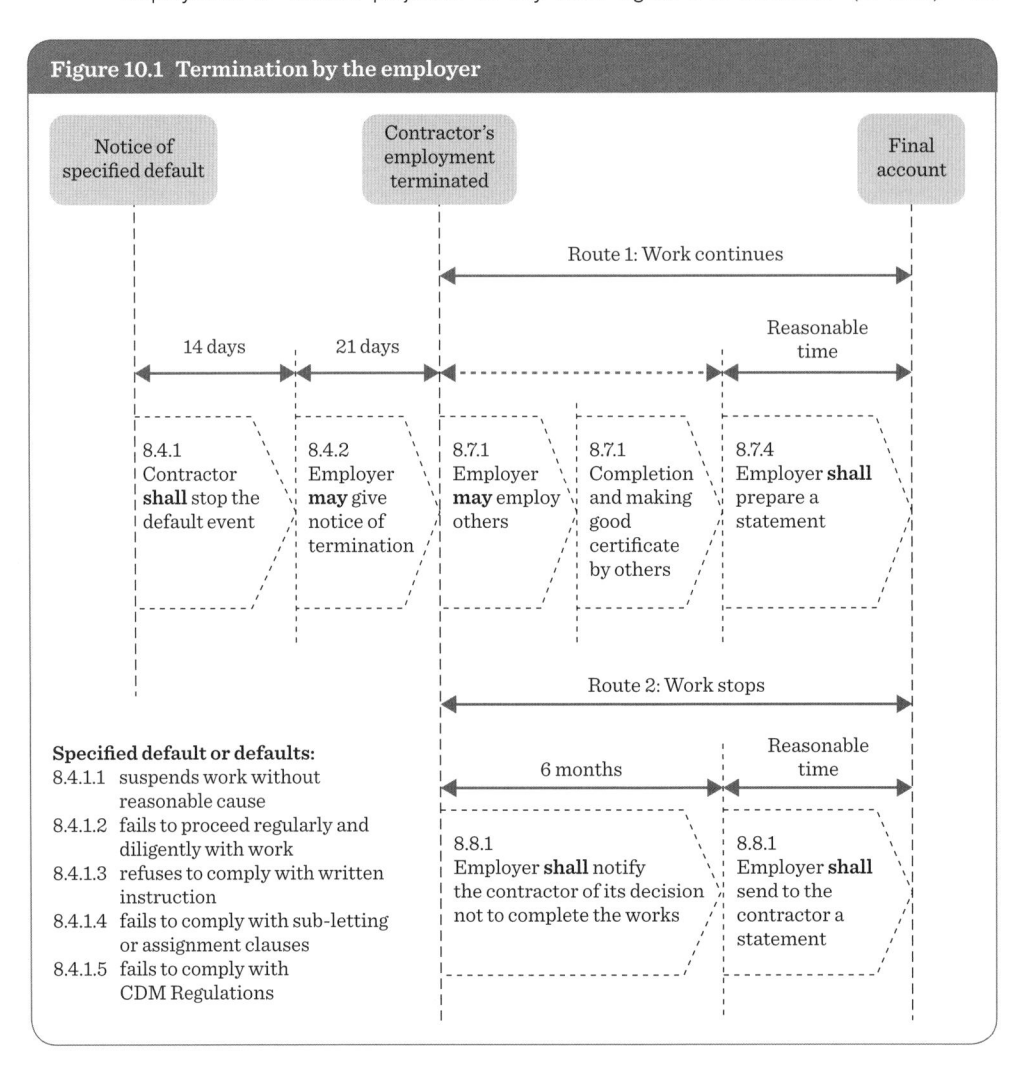

Figure 10.1 Termination by the employer

Notice of specified default — 14 days — 21 days — Contractor's employment terminated — Route 1: Work continues — Reasonable time — Final account

- 8.4.1 Contractor **shall** stop the default event
- 8.4.2 Employer **may** give notice of termination
- 8.7.1 Employer **may** employ others
- 8.7.1 Completion and making good certificate by others
- 8.7.4 Employer **shall** prepare a statement

Route 2: Work stops — 6 months — Reasonable time

- 8.8.1 Employer **shall** notify the contractor of its decision not to complete the works
- 8.8.1 Employer **shall** send to the contractor a statement

Specified default or defaults:
8.4.1.1 suspends work without reasonable cause
8.4.1.2 fails to proceed regularly and diligently with work
8.4.1.3 refuses to comply with written instruction
8.4.1.4 fails to comply with sub-letting or assignment clauses
8.4.1.5 fails to comply with CDM Regulations

termination can be initiated by the employer in the event of specified defaults by the contractor occurring prior to practical completion (cl 8.4.1), the insolvency of the contractor (cl 8.5) or corruption (cl 8.6). Where the employer is a local or public authority, circumstances as set out in regulation 73(1)(b) of the Public Contracts Regulations 2015 (conviction of various offences) will also give rise to the right to terminate (cl 8.6).

10.8   The procedures as set out in the contract must be followed exactly, especially those concerning the issue of notices (see Figure 10.1). If default occurs, the employer should issue a warning notice, which should specify the particular default (although not necessarily a detailed list of circumstances) (cl 8.4.1). If the default continues for 14 days from receipt of the notice, then the employer may terminate the employment of the contractor by the issue of a further notice within 21 days from the expiry of that 14-day period (cl 8.4.2). If the contractor ends the default or if the employer gives no further notice and the contractor then repeats the default, the employer may terminate 'within a reasonable time after such repetition' (cl 8.4.3). The employer must still give notice of termination, but no further warning is required. There appears to be no time limit on the repetition of the default. If, however, a considerable period has elapsed, it may be prudent for the employer to issue a further warning notice before issuing the notice of termination.

10.9   It should be noted that, to be valid, all notices must be given strictly in accordance with clause 1.7.4, i.e. 'by hand or sent by Recorded Signed for or Special Delivery post' (cl 8.2.3). It should be noted that this is not the same as the older wording 'actual delivery' so it is unlikely that fax or email would be acceptable, as it was in the case of *Construction Partnership* v *Leek Developments*. Notices sent by post are deemed to have been received 'on the second Business Day' after posting, unless there is proof to the contrary. As time limits are of vital importance here, it is usually wise to have receipt of delivery confirmed.

> *Construction Partnership UK Ltd* v *Leek Developments Ltd* [2006] CILL 2357 (TCC)
>
> On an IFC98 contract, a notice of determination was delivered by fax, but not by hand or by special delivery or recorded delivery. (A letter had been sent by normal post but it was unclear whether or not it had been received.) Clause 7.1 required actual delivery of notices of default and determination, and the contractor disputed whether the faxed notice was valid. The court had therefore to decide what 'actual delivery' meant. It decided that it meant what it says: 'Delivery simply means transmission by an appropriate means so that it is received'. In this case, it was agreed that the fax had been received, therefore the notice complied with the clause. The CILL editors state that 'on a practical level, this judgment is quite important' because it had previously been assumed that 'actual delivery' meant physical delivery by hand. In their view, email could be considered an appropriate method of delivery, although that was not decided in the case.

10.10   The grounds for termination by the employer must be clearly established and expressed. The contract clearly states that termination must not be exercised 'unreasonably or vexatiously' (cl 8.2.1). Under clause 8.4.1.1, suspension of the work must be whole and substantial, and 'without reasonable cause'. However, the contractor might find 'reasonable cause' in any of the matters referred to in clause 4.21. An exercise of the right to suspend work under clause 4.11 would not be cause for termination, provided that the right had been exercised in accordance with the terms of the contract.

10.11   The specified defaults which may give rise to termination are that the contractor:

- wholly or substantially suspends the design or construction of the works (cl 8.4.1.1);
- 'fails to proceed regularly and diligently with the performance of his obligations' (cl 8.4.1.2);
- refuses or neglects to comply with a written notice or instruction requiring the contractor to remove defective work (cl 8.4.1.3);
- fails to comply with clause 3.3 (sub-contracting) or clause 7.1 (assignment) (cl 8.4.1.4);
- fails to comply with clause 3.16 (CDM Regulations) (cl 8.4.1.5).

10.12   Generally speaking, only a serious default would justify termination, although any failure to comply with the CDM Regulations' provisions which would put the employer at risk of action by the authorities would be sufficient.

10.13   The default that the contractor 'fails to proceed regularly and diligently' (cl 8.4.1.2) is notoriously difficult to establish. Careful records kept by the employer or the employer's agent will be of utmost importance should the contractor decide to dispute this matter. 'Default' here means more than simply falling behind any submitted programme (relating to design and/or construction work), even to such an extent that it is quite clear the project will finish considerably behind time. However, something less than a complete cessation of design or construction work would be sufficient grounds.

10.14   In the case of *London Borough of Hounslow* v *Twickenham Garden Developments*, for example, an architect's notice of determination was strongly attacked by the defendant, although the facts should be contrasted with the more recent case of *West Faulkner Associates* v *London Borough of Newham*. The employer would be wise to proceed with considerable caution, and should bear in mind that, without the first 'warning notice', it has no right to terminate the contractor's employment.

> *London Borough of Hounslow* v *Twickenham Garden Developments* (1970) 7 BLR 81
>
> The London Borough of Hounslow entered into a contract with Twickenham Garden Developments to carry out sub-structure works at Heston and Isleworth in Middlesex. The contract was on JCT63. Due to a strike, work on the contract stopped for approximately eight months. After work resumed, the architects issued a notice of default stating that the contractor had failed to proceed regularly and diligently and that, unless there was an appreciable improvement, the contract would be determined. The employers then proceeded to determine the contractor's employment. The contractor disputed the validity of the notices and the determination and refused to stop work and leave the site. The Council applied to the court for an injunction to remove the contractor. The judge emphasised that an injunction was a serious remedy and that, before he could grant one, there had to be clear and indisputable evidence of the merits of the Council's case. The evidence put before him, which showed a significant drop in the amounts of monthly certificates and numbers of workers on site, failed to provide this.

> *West Faulkner Associates* v *London Borough of Newham* (1992) 61 BLR 81
>
> West Faulkner Associates were architects engaged by the London Borough of Newham for the refurbishment of a housing estate consisting of several blocks of flats. The residents of the

> estate were evacuated from their flats in stages to make way for the contractors, Moss, who, it had been agreed, would carry out the work according to a programme of phased possession and completion, with each block taking nine weeks. Moss fell behind the programme almost immediately. However, Moss had a large workforce on the site and continually promised to revise its programme and working methods to address the problems of lateness, poor quality work and unsafe working practices that were drawn to its attention on numerous occasions by the architects. In reality, Moss remained completely disorganised, and there was no apparent improvement. The architects took the advice of quantity surveyors that the grounds of failing to proceed regularly and diligently would be difficult to prove, and decided not to issue a notice. As a consequence, the Borough was unable to issue a notice of determination, had to negotiate a settlement with the contractor, and dismissed the architects, who then brought a claim for their fees. The judge decided that the architects were in breach of contract in failing to give proper consideration to the use of the determination provisions. In his judgment, he stated that '"regularly and diligently" should be construed together and in essence they mean simply that the contractors must go about their work in such a way as to achieve their contractual obligations. This requires them to plan their work, to lead and manage their workforce, to provide sufficient and proper materials and to employ competent tradesmen, so that the works are carried out to an acceptable standard and that all time, sequence and other provisions are fulfilled' (Judge Newey at page 139).

Insolvency of the contractor

10.15 Insolvency is the inability to pay debts as they become due for payment. Insolvent individuals may be declared bankrupt. Insolvent companies may be dealt with in a number of ways, depending on the circumstances: for example, by voluntary liquidation (in which the company resolves to wind itself up); compulsory liquidation (under which the company is wound up by a court order); administrative receivership (a procedure to assist the rescue of a company under appointed receivers); an administration order (a court order given in response to a petition, again with the aim of rescue rather than liquidation, and managed by an appointed receiver); or voluntary arrangement (in which the company agrees terms with creditors over payment of debts). Procedures for dealing with insolvency are mainly subject to the Insolvency Act 1986 and the Insolvency Rules 1986 (SI 1986/1925). Under these provisions, the person authorised to oversee statutory insolvency procedures is termed an 'insolvency practitioner'.

10.16 Under DB16, the contractor must notify the employer in writing in the event of liquidation or insolvency (cl 8.5.2). A contractual definition of insolvency is given in clause 8.1. Termination is not automatic, however, and this allows the appointed insolvency practitioner time to come up with a rescue package, if one is possible. It is usually in the employer's interest to have the works completed with as little additional delay and cost as possible, and a breathing space may allow all possibilities to be explored. During this period, the contract states that 'clauses 8.7.3 to 8.7.5 and (if relevant) clause 8.8 shall apply as if such notice had been given' (cl 8.5.3.1). This means that even if no notice of termination is given, the employer is under no obligation to make further payment except as provided under those clauses (see paragraph 10.23). The contractor is relieved of the obligation to 'carry out and complete the Works' (cl 8.5.3.2). The employer may then take reasonable steps to ensure that the site, works and materials are secure and protected, and the contractor must not hinder such measures (cl 8.5.3.3).

10.17 There are several options for completing the works. The first is for arrangements to be made so that the contractor may continue with the work. Unless the insolvency practitioner has been able to arrange resource backing, this may not be a realistic option. If practical completion is near, however, and money is due to the contractor, it may be advantageous to allow completion under the control of the insolvency practitioner.

10.18   Alternatively, another contractor could be novated to complete the works. On a 'true novation', the substitute contractor will take over all the original obligations and benefits (including completion to time and within the contract sum). A more likely arrangement would be a 'conditional novation', whereby the contract completion date, etc. would be subject to renegotiation, and the substitute contractor would probably want to disclaim liability for that part of the work undertaken by the original contractor.

10.19   Deciding which of the options would best serve the interests of all the parties is a matter to be resolved between the employer, the insolvency practitioner and the contractor. A pragmatic approach to the issue of completion may be to continue initially with the original contractor under an interim arrangement until such time as novation can be arranged or a completion contract negotiated.

Consequences of termination by the employer

10.20   If the employer exercises its right to terminate under clause 8.4, 8.5 or 8.6, then completion will only be achieved through the appointment of a new contractor of the employer's choice. Clause 8.7.1 gives the employer the right to employ others to complete the works and completion would include making good any defects in the work already carried out and completing the design. A completion contract might result from negotiation or competitive tender. The employer will have the right to use any temporary buildings, plant, etc. on site, including those which are not owned by the original contractor, subject to the consent of the owner (cl 8.7.1).

10.21   The employer may also require the contractor to:

- remove from the site any temporary buildings, plant, etc. which are owned by the contractor (cl 8.7.2.1);
- provide the employer with copies of all contractor's design documents (cl 8.7.2.2);
- require the original contractor to assign the benefit of any sub-contracts to the employer (to the extent that the benefit is assignable) (cl 8.7.2.3).

10.22   If the employer decides to employ others under clause 8.7.1, such employment must be handled with care, as completion of a building started by another contractor is always difficult. A completion contract might result from negotiation or competitive tender, but the tender route may be advisable if there is much to complete as the employer may have to demonstrate subsequently that the costs incurred were reasonable. A record should be made of the exact state of completeness at the time of termination, including any defective work.

10.23   Following termination, clause 8.7.3 states that 'no further sum shall become due to the Contractor under this Contract other than any amount that may become due to him under clause 8.7.5 or 8.8.2' (i.e. the payment provisions following termination). It also states that the employer will not need to make any payments that have already become due to the extent that a pay less notice has been given (cl 8.7.3.1) or where the contractor has become insolvent after the final date when a pay less notice could have been issued (cl 8.7.3.2). This reflects section 111(10) of the HGCRA 1996 as amended and the judgment in *Melville Dundas* v *George Wimpey*. Therefore, if the contractor becomes insolvent after a certificate is issued, but before the final date for issuing a pay less notice, then to avoid any disputes the employer should issue a notice. It should be noted, however, that following a termination the employer may still be obliged to pay amounts awarded by an adjudicator (*Ferson Contractors* v *Levolux*).

> *Melville Dundas Ltd* v *George Wimpey UK Ltd* [2007] 1 WLR 1136 (HL)
>
> On a contract let on WCD98, the contractor had gone into receivership, entitling the employer to determine the contractor's employment. The contractor had applied for an interim payment on 2 May 2003, the final date for payment was 16 May (14 days after application) and the determination was effective on 30 May 2003. The contractor claimed the payment on the basis that no withholding notice had been issued. By a majority of three to two, the House of Lords decided that the employer was not obliged to make any further payment. It was accepted that, under WCD98, interim payments were not contractually payable after determination and the House of Lords held that this was not inconsistent with the payment provisions of the HGCRA 1996. Although the Act requires that the contractor should be entitled to payment in the absence of a notice, this did not mean that that entitlement had to be maintained after the contractor had become insolvent, i.e. it was not inconsistent to construe that the effect of the determination was that the payment was no longer due. The Act was concerned with the balance of interests between payer and payee, and to construe it otherwise would give a benefit to the contractor's creditors against the interests of the employer, something which the Act did not intend.

> *Ferson Contractors Ltd* v *Levolux A T Ltd* [2003] BLR 118
>
> Ferson was the contractor and Levolux the sub-contractor on a GC/Works sub-contract. A dispute arose regarding Levolux's second application for payment; £56,413 was claimed but only £4,753 was paid. A withholding notice was issued, which specified the amount, but not the reason for withholding it. Levolux brought a claim to adjudication, and the adjudicator decided that the notice did not comply with section 111 of the HGCRA 1996, and that Ferson should pay the whole amount. Ferson refused to pay and Levolux sought enforcement of the decision. Prior to the adjudication, Levolux had suspended work and Ferson, maintaining that the suspension was unlawful, had determined the contract. It now maintained that, due to clause 29, which stated that 'all sums of money that may be due or accruing due from the contractor's side to the sub-contractors shall cease to be due or accrue due', they did not have to pay this amount. The CA upheld the decision of the judge of first instance that the amount should be paid: 'The contract must be construed so as to give effect to the intent of Parliament'.

10.24 Following termination a notional final account must be set out, stating what is owed or owing, in a statement prepared by the employer (cl 8.7.4). This account must be prepared within three months of 'the completion of the Works and the making good of defects in them', which allows the employer a period to assess its losses due to the termination. The net amount shown on the account should be paid by the contractor to the employer (the more likely outcome), or by the employer to the contractor, as appropriate (cl 8.7.5).

10.25 One consequence of termination is that it often takes time for the contractor to effect an orderly withdrawal from site, and for the employer to establish the amounts outstanding before final payment. Should the employer decide not to continue with the construction of the works after termination, the employer is required to notify the contractor in writing within six months of termination (cl 8.8.1). Within a reasonable time following notification (or within six months of termination, if no work is carried out and no notice issued), the employer must send the contractor a statement of the value of the works and losses suffered within two months of the expiry of the six-month period, as required under clause 8.8.1.

Termination by the contractor

10.26 The contractor has a reciprocal right to terminate its own employment under clause 8.9 in the event of specified defaults of the employer (cl 8.9.1) or specified suspension events (cl 8.9.2), or insolvency of the employer (cl 8.10) (see Figure 10.2). The specified events must have resulted in the suspension of the whole of the uncompleted works for the continuous period stated in the contract particulars. In the case of specified defaults or suspension events a notice is required, which must specify the default or event. If the default or event continues for 14 days from receipt of the notice, the contractor may terminate the employment by a further notice up to 21 days from the expiry of the 14 days. Alternatively, if the employer ends the default or the suspension event ceases, and the contractor gives no further notice, then, should the employer repeat the default, the contractor may terminate 'within a reasonable time after such repetition' (cl 8.9.4). As for the employer, these notices must be given by the means set out in clause 1.7.4, i.e. 'by hand or sent by Recorded Signed for or Special Delivery post' (cl 8.2.3).

10.27 The grounds of clause 8.9 differ from those that give the employer the right to terminate. They include failure to pay an amount properly due, and failure to comply with the contractual provisions relating to assignment under clause 7.1 or provisions relating to the

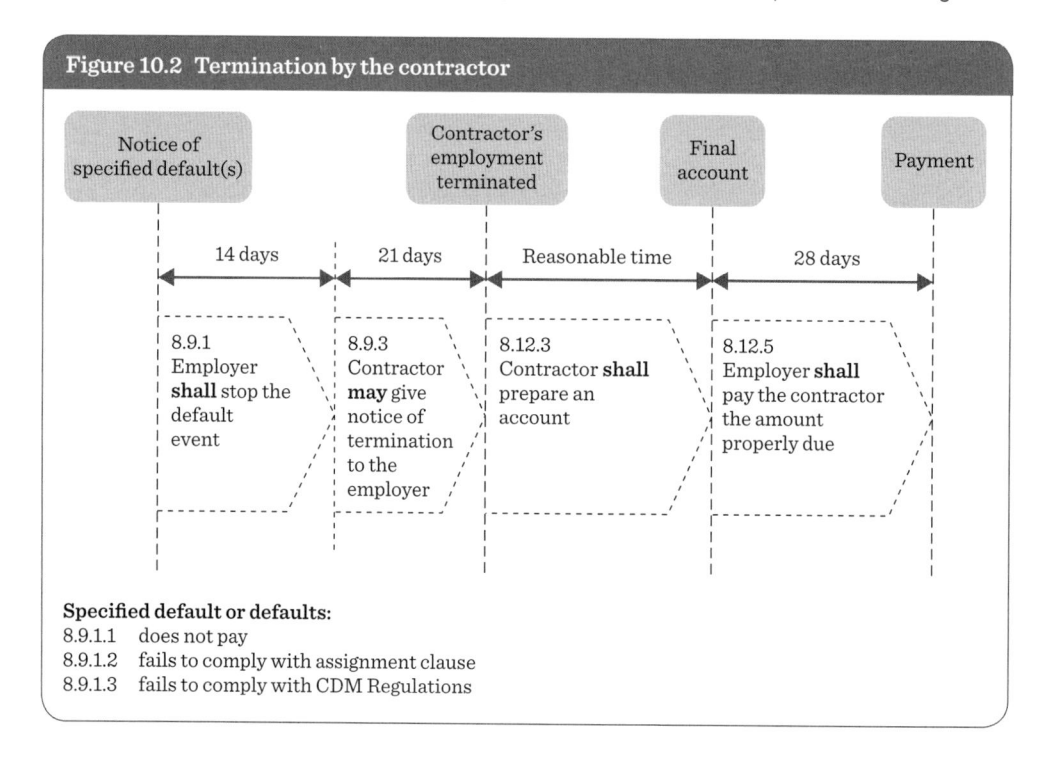

Figure 10.2 Termination by the contractor

Notice of specified default(s) — 14 days — Contractor's employment terminated — 21 days — Final account — Reasonable time — Payment — 28 days

8.9.1 Employer **shall** stop the default event

8.9.3 Contractor **may** give notice of termination to the employer

8.12.3 Contractor **shall** prepare an account

8.12.5 Employer **shall** pay the contractor the amount properly due

Specified default or defaults:
8.9.1.1 does not pay
8.9.1.2 fails to comply with assignment clause
8.9.1.3 fails to comply with CDM Regulations

CDM Regulations under clause 3.16. In addition, the contractor would have grounds where the carrying out of the whole or substantially the whole of the works is suspended for a period of one month (or any period stated in the contract particulars) due to 'any impediment, prevention or default, whether by act or omission, by the Employer or any Employer's Person' (cl 8.9.2), unless it is necessitated by some negligence or default of the contractor.

10.28 The contractor should exercise particular care if considering termination due to what it considers to be an employer's failure to pay. The contract specifically requires failure to pay 'an amount properly due', and not simply the amount for which the contractor might have applied. If the contractor has made an error in its calculations, the employer might be entitled to pay a lesser amount. In addition, the contract gives the employer rights to make various adjustments to and deductions from amounts due. The correct exercise of these rights would not amount to a failure to pay an amount properly due. If the contractor attempts to terminate the contract without justification, this will amount to repudiation, with serious consequences for the contractor. A more prudent course would be to raise the disputed payment in adjudication, while continuing to proceed with the works.

10.29 Termination by the contractor is optional in the case of the employer's bankruptcy or insolvency (cl 8.10.1). The contractor must issue a notice and termination will take effect from the receipt of the notice.

Consequences of termination by the contractor

10.30 In the event of termination, the contractor must provide the employer with copies of all drawings and other information completed prior to termination (cl 8.12.2.2). It must also remove from the site all temporary buildings, tools, etc. (cl 8.12.2.1). The contractor should then prepare an account setting out the total value of the work at the date of termination, including the cost of design work and other costs relating to the termination as set out in clause 8.12.3. These may include such items as the cost of removal and any direct loss and/or damage consequent upon termination (cl 8.12.3.3 and 8.12.3.5). The contractor is, in effect, indemnified against any damages that may be caused as a result of the termination. This would not necessarily be the case if the contractor did not comply with the contractual provisions; such non-compliance might constitute repudiation.

Termination by either the employer or the contractor

10.31 Either party has the right to terminate the contract if the carrying out of the whole (or substantially the whole) of the works is suspended for the period inserted in the contract particulars (if none is stated, this is two months) because of force majeure, loss or damage to the works caused by any risk covered by the works insurance policy or by an excepted risk, civil commotion, the exercise by the government or a local or public authority of a statutory power, or delay in the receipt of statutory approvals or permissions which the contractor has taken all practicable steps to avoid (cl 8.11.1). The right of the contractor to terminate in the event of damage to the works is limited by the proviso that the event must not have been caused by the contractor's negligence (cl 8.11.2). In addition, either party may terminate if work is suspended because of an employer's instruction under clause 2.13 (discrepancies), clause 3.9 (changes) or clause 3.10 (postponement) which has been issued as a result of negligence or default of a statutory undertaker (cl 8.11.1.2).

Figure 10.3 Termination by either party

| Suspension of works | Warning notice from either party | | Contractor's employment terminated | Final account | Payment |

Timeline:
- 2 months (or any stated in contract particulars)
- 7 days
- Reasonable time
- 28 days

- 8.11.1 Uncompleted works are suspended
- 8.11.1 Either party may give notice of termination
- 8.11.1 Either party may terminate contractor's employment
- 8.12.3 Contractor **shall** prepare an account
- 8.12.5 Employer **shall** pay the contractor the amount properly due

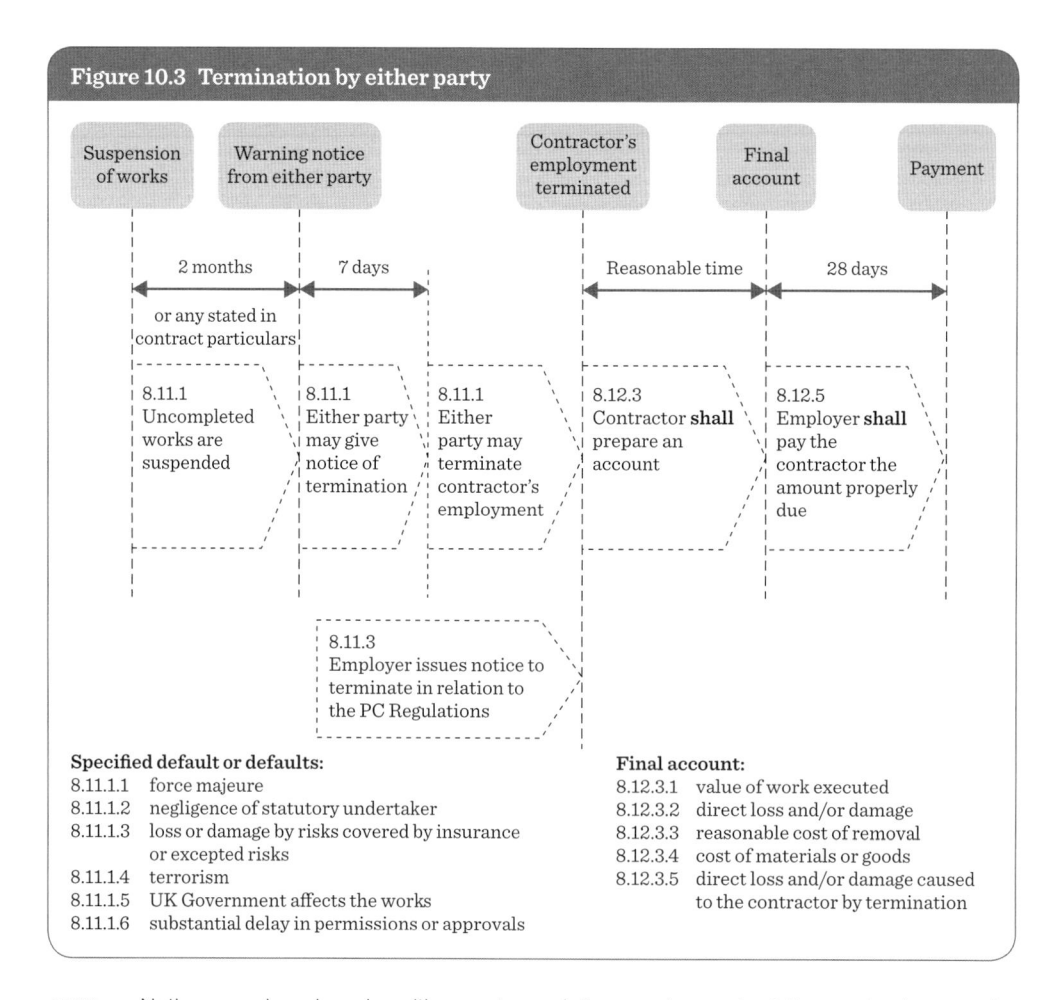

- 8.11.3 Employer issues notice to terminate in relation to the PC Regulations

Specified default or defaults:
- 8.11.1.1 force majeure
- 8.11.1.2 negligence of statutory undertaker
- 8.11.1.3 loss or damage by risks covered by insurance or excepted risks
- 8.11.1.4 terrorism
- 8.11.1.5 UK Government affects the works
- 8.11.1.6 substantial delay in permissions or approvals

Final account:
- 8.12.3.1 value of work executed
- 8.12.3.2 direct loss and/or damage
- 8.12.3.3 reasonable cost of removal
- 8.12.3.4 cost of materials or goods
- 8.12.3.5 direct loss and/or damage caused to the contractor by termination

10.32 Notice may be given by either party, and the employment of the contractor may be terminated seven days after receipt of the notice, unless the suspension is terminated within that period. If work is not resumed after this period, the party may then, by further notice, terminate the contract (cl 8.11.1, see Figure 10.3). Where the employer is a local or public authority, the employer may issue a notice if circumstances in regulation 73(1)(a) or (c) of the Public Contracts Regulations 2015 (various breaches of the Regulations) apply (cl 8.11.3).

10.33 Detailed provisions are set out regarding the consequences of the termination. Clause 8.12.1 states that 'no further sums shall become due to the Contractor otherwise than in accordance with this clause', which in effect means that other provisions of the contract requiring further payment will cease to operate. The contractor must remove all temporary buildings, tools, etc. from the site (cl 8.12.2). An account is then prepared in the same format as for termination by the contractor (cl 8.12.3, see paragraph 10.30 above), except that, in this case, it may include amounts relating to direct loss and/or damage to the contractor resulting from a specified peril caused by the employer's negligence.

Termination of named sub-contractor's employment

10.34   Where a named sub-contractor has been appointed under Supplemental Provision 1, the contractor may only terminate its employment with the prior consent of the employer, except in cases where the sub-contractor has become insolvent (Schedule 2:1.3.2). The contractor must first notify the employer of any intended termination, and then, if the employer consents, may terminate the contract, sending copies of the notices to the employer (Schedule 2:1.3.1 and 1.3.3). If such termination takes place, the contractor must complete any outstanding work (Schedule 2:1.4.1), and this shall be 'treated as a Change', except where the termination results from some default of the contractor or where the contractor has not obtained the prior consent of the employer (Schedule 2:1.4.2). For both exceptions, the contractor would have to complete the work at its own risk. Otherwise, the work would be valued under clause 5.2, and it may provide grounds for an extension of time and for a claim for loss and/or expense (cl 2.26.1 and 4.21.1).

10.35   The contractor is required to account for any amounts it recovers (or could have recovered) from the named sub-contractor using reasonable diligence (Schedule 2:1.4.3). This means that any additional amounts payable to the contractor as a result of the change would be reduced by any amounts the contractor could have recovered from the sub-contractor.

10.36   As there are few restrictions on the terms under which the named sub-contractor is engaged, the sums that could be recovered may amount to very little. Some protection is afforded to the employer, however, by the requirement that the contractor includes the term set out in Schedule 2 paragraph 1.5 in any agreement with a named sub-contractor. This requirement is one of the more obscure examples of JCT drafting, but it is, nevertheless, an important clause. By this provision the named sub-contractor promises not to contend that the contractor has suffered no losses or that its liability to the contractor is reduced, due to the operation of the clauses in Supplemental Provision 1. The concern is that Supplemental Provision 1 has created a trap for the employer. As the contractor will, in due course, be compensated under Schedule 2 paragraph 1.4.2 for its costs in completing the work, the contractor will, in effect, suffer no losses. The sub-contractor might therefore claim that it is under no obligation to compensate the contractor. If this argument were to succeed, the contractor would not recover anything, and so would be unable to pass on any compensation to the employer. The Schedule 2 paragraph 1.5 provision is intended to break this vicious circle by preventing the sub-contractor from using this argument. This allows the contractor to claim losses from the sub-contractor, which it can then pass on to the employer.

11 Dispute resolution

11.1 DB16 refers to five methods of dispute resolution: negotiation; mediation; adjudication; arbitration; and legal proceedings. One of these methods, adjudication, is a statutory right, and if one party wishes to use this method, the other must concur. Negotiation is an optional provision (Supplemental Provision 10). Negotiation and mediation are voluntary processes which depend on the co-operation of the parties, and either may lead to a binding result. If none of the options of negotiation, mediation or adjudication is used, or if either party is dissatisfied with the decision of an adjudicator, then the dispute will have to be resolved by arbitration or litigation.

11.2 DB16 requires the parties to decide in advance whether arbitration or litigation will be used. If arbitration is to be the final method of dispute resolution, then the contract particulars must indicate that 'Article 8 and clauses 9.3 to 9.8 apply'. If this is the case, then all disputes will be referred to arbitration, except for those relating to the Construction Industry Scheme, VAT or the enforcement of an adjudicator's decision.

11.3 There are therefore stages, either before or during the contract, where the parties have the opportunity to agree a preferred course of action, so it is important for them to understand the alternative methods.

Notification and negotiation

11.4 If Supplemental Provision 10 is incorporated, the parties are each obliged to notify the other promptly of any matter that may give rise to a dispute. The senior executives nominated in the contract (or persons of equal standing) must then meet and, in good faith, try to resolve the matter. There is no sanction for non-compliance, but refusal or failure to comply might be taken into account in any subsequent legal proceedings.

Mediation

11.5 If negotiations fail to achieve an agreement, the parties may wish to submit the dispute to 'alternative dispute resolution' (ADR), a term used to cover methods such as conciliation, mediation and the mini-trial. DB16 clause 9.1 requires each party to give serious consideration to a request by the other to use mediation. Footnote [57] to clause 9.1 refers to the Guide (DB/G), although the Guide does not in fact give any detailed explanation of this process. (The DB16 guide states that such choices are frequently better made by the parties when the dispute has actually arisen.) The parties could, of course, supplement DB16 by selecting a mediator or mediator-appointing body and setting this out in their contract. The RIBA, RICS, ACIArb and CEDR all maintain lists of mediators and will give guidance on appointment and procedure.

11.6 As mediation is a consensual process, any referral of a dispute to mediation would have to be supported by both parties. Usually a mediator is appointed jointly by the parties, and will normally meet with the parties together and separately in an attempt to resolve the differences. The outcome is in the form of a recommendation which, if acceptable, can be signed as a legally binding agreement, enforceable in the same way as any other contract. However, if the recommendation is not acceptable to one of the parties and is not signed as a binding agreement, it cannot be imposed by law, and so the time spent on the mediation may appear to have been wasted.

11.7 Nevertheless, there can be many advantages to mediation. Unlike adjudication, arbitration or litigation, it is a non-adversarial process which tends to forge good relationships between the parties. Imposed solutions may leave at least one of the parties dissatisfied and may make it very difficult for the parties to work together in the future. If the parties seek to promote a long-term business relationship, then mediation merits serious consideration. Even if mediation does not result in a complete solution, it has been found in practice that it can help to clear the air on some of the issues involved and to establish common ground. This, in turn, might then pave the way for shorter and possibly less acrimonious arbitration or litigation.

Adjudication

11.8 The Housing Grants, Construction and Regeneration Act (HGCRA) 1996 Part II as amended by the Local Democracy, Economic Development and Construction Act (LDEDCA) 2009 requires that parties to construction contracts falling within the definition set out in the Act have the right to refer any dispute arising 'under the contract' to a process of adjudication which complies with requirements stipulated in the Act. Article 7 of DB16 restates this right, and refers to clause 9.2, which states that where a party decides to exercise this right 'the Scheme shall apply'. This refers to the Scheme for Construction Contracts 1998 (as amended in 2011), a piece of secondary legislation which sets out a procedure for the appointment of the adjudicator and the conduct of the adjudication. The Scheme takes effect as implied terms in a contract, if and to the extent that the parties have failed to agree on a procedure that complies with the Act.

11.9 By stating 'the Scheme shall apply', DB16 effectively annexes the provisions of the Scheme to the form, which therefore become a binding part of the agreement between the parties. Clause 9.2, however, makes its application subject to certain conditions which relate to the appointment of the adjudicator.

11.10 Under DB16, the adjudicator may be either named in the contract particulars or nominated by the nominating body identified in the contract particulars. It should be noted that the list of bodies has changed since DB05, and now includes, as well as the RIBA, the RICS and the CIArb, the 'constructionadjudicators.com' and the Association of Independent Construction Adjudicators (AICA). A named adjudicator will normally enter into the JCT Adjudication Agreement Named Adjudicator (Adj/N) with the parties at the same time as the main contract is entered into.

11.11 The party wishing to refer the dispute to adjudication must give notice under paragraph 1(1) of the Scheme. The notice may be issued at any time and should identify briefly the dispute or difference, give details of where and when it has arisen, set out the nature of the redress sought, and include the names and addresses of the parties, including any specified for the giving of notices (paragraph 1(3)). If no adjudicator is named, the parties may either agree an adjudicator or either party may apply to the nominating body identified in the contract particulars (paragraph 2(1)). If no nominating body has been selected, then the contract states that the referring party may apply to any of the bodies listed in the contract particulars. The adjudicator will then send terms of appointment to the parties. In addition to the form for a named adjudicator, the JCT also publishes an Adjudication Agreement (Adj) for use in this situation.

11.12 The Scheme does not stipulate any qualifications in order to be an adjudicator, but does state that the adjudicator 'should be a natural person acting in his personal capacity' and should not be an employee of either of the parties (paragraph 4). In addition, DB16 requires that, where the dispute relates to clause 3.13.3 (repeat testing), the person appointed shall 'where practicable' be 'an individual with appropriate expertise and experience in the specialist area or discipline relevant to the instruction or issue in dispute' (cl 9.2.2.1). Where the person does not have the appropriate expertise, he or she must appoint an independent expert to advise and report.

11.13 The referring party must refer the dispute to the selected adjudicator within seven days of the date of the notice (paragraph 7(1)). The referral will normally include particulars of the dispute, and must include a copy of, or relevant extracts from, the contract, and any material it wishes the adjudicator to consider (paragraph 7(2)). A copy of the referral must be sent to the other party and the adjudicator must inform all parties of the date it was received (paragraph 7(3)).

11.14 The adjudicator will then set out the procedure to be followed. A preliminary meeting may be held to discuss this, otherwise the adjudicator may send the procedure and timetable to both parties. The party which did not initiate the adjudication (the responding party) will be required to respond by a stipulated deadline. The adjudicator is likely to hold a short hearing of a few days at which the parties can put forward further arguments and evidence. There may also be a site visit. Sometimes it may be possible to carry out the whole process by correspondence (often termed 'documents only').

11.15 The adjudicator is given considerable powers under the Scheme (e.g. paragraphs 13 and 20), including the right to take the initiative in obtaining the facts and the law, the right to issue directions, the right to revise decisions and certificates of the contract administrator, the right to carry out tests (subject to obtaining necessary consents) and the right to obtain from others necessary information and advice. The adjudicator must give advance notice if intending to take legal or technical advice.

11.16 The HGCRA 1996 requires that the decision is reached within 28 days of referral, but it does not state how this date is to be established (section 108(2)(c)). Under the Scheme, the 28 days start to run from the date of receipt of the referral notice (section 19(1)). The period can be extended by up to 14 days by the referring party, and further by agreement between the parties. The decision must be delivered forthwith to the parties, and the adjudicator may not retain it pending payment of the fee. The provisions state that the adjudicator must give reasons for the decision if requested to do so by the parties (section 22).

11.17 The parties must meet their own costs of the adjudication, unless they have agreed that the adjudicator shall have the power to award costs. Under the Act, any agreement is ineffective unless it complies with section 108A, including that it is made in writing after a notice of adjudication is issued (DB16 therefore does not contain such an agreement). The adjudicator, however, is entitled to charge fees and expenses (subject to any agreement to the contrary), although expenses are limited to those 'reasonably incurred' (paragraph 25). The adjudicator may apportion those fees between the parties, and the parties are jointly and severally liable to the adjudicator for any sum which remains outstanding following the adjudicator's determination. This means that in the event of default by one party, the other party becomes liable to the adjudicator for the outstanding amount.

11.18 The adjudicator's decision will be final and binding on the parties 'until the dispute is finally determined by legal proceedings, by arbitration, or by agreement between the parties'. The effect of this is that if either party is dissatisfied with the decision, it may raise the dispute again in arbitration or litigation as indicated in the contract particulars, or it may negotiate a fresh agreement with the other party. In all cases, however, the parties remain bound by the decision and must comply with it until the final outcome is determined.

11.19 If either party refuses to comply with the decision, the other may seek to enforce it through the courts. Generally, actions regarding adjudicators' decisions have been dealt with promptly by the courts and the recalcitrant party has been required to comply. Paragraph 22A of the Scheme allows the adjudicator to correct clerical or typographical errors in the decision, within five days of it being issued, either on its own initiative or because the parties have requested it, but this would not extend to reconsidering the substance of the dispute.

Arbitration

11.20 Arbitration refers to proceedings in which the arbitrator has power derived from a written agreement between the parties to a contract, and which is subject to the provisions of the Arbitration Act 1996. Arbitration awards are enforceable at law. An arbitrator's award can be subject to appeal on limited grounds.

11.21 If arbitration is preferred to litigation as the method for final resolution of disputes, then this is confirmed by selecting Article 8 in the contract particulars. The range of disputes that can be referred to arbitration is wider than for adjudication, as Article 8 refers to 'any dispute or difference between the Parties of any kind whatsoever arising out of or in connection with this Contract'. The arbitration provisions are set out in clauses 9.3 to 9.8, which refer to the Construction Industry Model Arbitration Rules (the Rules). The Arbitration Act 1996 confers wide powers on the arbitrator unless the parties have agreed otherwise, but leaves detailed procedural matters to be agreed between the parties or, if not so agreed, to be decided by the arbitrator. To avoid problems arising, it is advisable to agree as much as possible of the procedural matters in advance, and DB16 does this by incorporating the Rules, which are very clearly written and self-explanatory. The specific edition referred to is the 2016 edition published by the JCT, which incorporates supplementary and advisory procedures, some of which are mandatory (Part A) and some of which apply only if agreed after the arbitration is begun (Part B). The paragraphs below refer to the JCT edition.

11.22 The party wishing to refer the dispute to arbitration must give notice as required by DB16 clause 9.4.1 and Rule 2.1, briefly identifying the dispute and requiring the party to agree to the appointment of an arbitrator. If the parties fail to agree within 14 days, either party may apply to the 'appointor', selected in advance from a list of organisations set out in the contract particulars. If no appointor is selected, then the contract states that the appointor will be the president or a vice-president of the RIBA. Under Rule 2.5, the arbitrator's appointment takes effect when they agree to act, and is not subject to reaching agreement with the parties on matters such as fees.

11.23   The arbitrator has the right and the duty to decide all procedural matters, subject to the parties' right to agree any matter (Rule 5.1). Within 14 days of appointment, the parties must each send to the arbitrator, and to each other, a note indicating the nature of the dispute and amounts in issue, the estimated length of the hearing, if necessary, and the procedures to be followed (Rule 6.2 and mandatory procedure 6.2.1). The arbitrator must hold a preliminary meeting within 21 days of appointment to discuss these matters (Rule 6.3 and mandatory procedure 6.3.1). The first decision to make is whether Rule 7 (short hearing), Rule 8 (documents only) or Rule 9 (full procedure) is to apply. If the parties cannot agree on a rule, mandatory procedure 6.3.2 states that Rule 8 shall apply, unless the arbitrator directs that Rule 9 shall apply. The decision will depend on the scale and type of dispute.

11.24   Under all three Rules referred to above, the parties exchange statements of claim and of defence, together with copies of documents and witness statements on which they intend to rely. Under Rule 8, the arbitrator makes the award based on the documentary evidence only. Under Rule 9, the arbitrator will hold a hearing at which the parties or their representatives can put forward further arguments and evidence. There may also be a site visit. The JCT amendments set out time limits for these procedures.

11.25   Under Rule 7, a hearing is to be held within 21 days of the date when Rule 7 becomes applicable, and the parties must exchange documents not later than seven days prior to the hearing. The hearing should last no longer than one day. The arbitrator publishes the award within one month of the hearing and the parties bear their own costs.

11.26   The arbitrator is given a wide range of powers under Rule 4, including:

- the power to obtain advice (Rule 4.2);
- the powers set out in section 38 of the Arbitration Act 1996 (Rule 4.3);
- the power to order the preservation of work, goods and materials, even though they are a part of work that is continuing (Rule 4.4);
- the power to request the parties to carry out tests (Rule 4.5);
- the power to award costs.

11.27   Under clause 9.5 of DB16, the arbitrator is also given wide powers to review and revise any certificate, opinion, decision, requirement or notice and to disregard them if need be, where seeking to determine all matters in dispute.

11.28   Where the arbitrator has the power to award costs, this will normally be done on a judicial basis, i.e. the loser will pay the winner's costs (Rule 13.1). The arbitrator is entitled to charge fees and expenses and will apportion those fees between the parties on the same basis. The parties are jointly and severally liable to the arbitrator for fees and expenses.

Arbitration and adjudication

11.29   Under Article 8, any dispute that has been referred to an adjudicator may be referred to arbitration if this is required by either party. Clause 1.8.2.1 states that, even where the decision has been given after the due date for the final payment, either party may refer the dispute to arbitration, provided the arbitration is commenced within 28 days of the adjudicator's decision.

Arbitration or litigation

11.30   As stated above, DB16 contains alternative provisions for arbitration and litigation in Articles 8 and 9, and a choice has to be made before tender documents are sent out. Both processes give rise to binding and enforceable decisions and both tend to be lengthy and expensive, although there are provisions for short forms of arbitration.

11.31   Litigation cases involving claims for amounts greater than £25,000 are normally heard in the High Court, and construction cases are usually heard in the Technology and Construction Court, a specialist department of the High Court which deals with technical or scientific cases. Procedures in court follow the Civil Procedure Rules, with the timetable and other detailed arrangements being determined by the court. A judge will hear the case and although, in the past, parties were required to be represented by barristers, now they may represent themselves or elect to be represented by an 'advisor'.

11.32   Disputes in building contracts have traditionally been settled by arbitration. Arbitrators are usually senior and experienced members of one of the construction professions and, for many years, it was felt that they had a greater understanding of construction projects and the disputes that arise than might be found in the courts. These days, however, the judges of the Technology and Construction Court have extensive experience of technical construction disputes. The high standards now evident in these courts are likely to be matched, in practice, by only a few arbitrators.

11.33   The court has powers to order that actions regarding related matters are joined (for example, where disputes between an employer and contractor, and contractor and sub-contractor, concern the same issues). This is much more difficult to achieve in arbitration. Even if all parties have agreed to the Construction Industry Model Arbitration Rules, the appointing bodies must have been alerted and have agreed to appoint the same arbitrator (Rules 2.6 and 2.7). If the same arbitrator is appointed, they may order concurrent hearings (Rule 3.7), but may only order consolidated proceedings with the consent of all the parties (Rule 3.9), which is often difficult to obtain. The court's powers may therefore offer an advantage in multi-party disputes, by avoiding duplication of hearings and possible conflicting outcomes.

11.34   There remain, however, two key advantages to using arbitration. The first is that in arbitration the proceedings can be kept private, which is usually of paramount importance to construction professionals and companies, and is often a deciding factor in selecting this process. In court, the proceedings are open to the public and the press, and the judgment is published and widely available.

11.35   The second advantage to the parties is that the arbitration process is consensual. The parties are free to agree on timing, place, representation and the individual arbitrator. This autonomy carries with it the benefits of increased convenience, and possible savings in time and expense. The parties avoid having to wait their turn at the High Court, and may choose a time and place for the hearing which is convenient to all. In arbitration, however, the parties have to pay the arbitrator and meet the cost of renting the premises in which the hearing is held.

11.36   It should perhaps be noted that, even where parties have selected arbitration under Article 8, it is still open for them to select litigation once a dispute develops. If, however, one party commences court proceedings, the other may ask the court to stay the proceedings on the grounds that an arbitration agreement already exists. This would not apply to litigation to enforce an adjudicator's decision, as Article 8 excludes all disputes regarding the enforcement of a decision of an adjudicator from the jurisdiction of the arbitrator. If, on the other hand, the parties had originally selected litigation, this would not prevent them from subsequently agreeing to take a dispute to arbitration, but, in such cases, they would also have to agree how the arbitrator is to be appointed and which procedural rules are to apply.

References

Publications

Aeberli, P. *Focus on Construction Contract Formation*, RIBA Publishing, London (2003)
Chappell, D. *The JCT Minor Works Building Contracts 2005*, Blackwell Publishing, Oxford (2006)
Chappell, D. *The JCT Design and Build Contract 2011*, Blackwell Publishing, Oxford (2014)
Construction Industry Council. *Building Information Model (BIM) Protocol* (CIC/BIM Pro), CIC, London (2013)
Construction Industry Council Liability Panel. *CIC Risk Management Briefing: Net Contribution Clauses*, London, CIC (2016)
Fletcher, P. and Satchwell, H. *Briefing: A Practical Guide to RIBA Plan of Work 2013 Stages 7, 0 and 1*, RIBA Publishing, London (2015)
Furst, S. and Ramsey, V. (eds.) *Keating on Construction Contracts*, 10th edn, London, Sweet & Maxwell (2016)
Hyams, D. *Construction Companion to Briefing*, RIBA Publishing, London (2001)
JCT. *Tendering Practice Note 2012*. Sweet & Maxwell, London (2012)
JCT. *Design and Build Contract Guide* (DB/G). Sweet & Maxwell, London (2016)
JCT and BDP. *JCT Guide to the Use of Performance Specifications*, RIBA Publishing, London (2001)
Lupton, S. *Cornes and Lupton's Design Liability in Construction*, 5th edition, Wiley-Blackwell, Chichester (2013)
NBS. *National Construction Contracts and Law Survey 2015*, NBS, Newcastle-upon-Tyne (2015)
RIBA. *Guide to RIBA Agreements 2010 (2012 revision)*, RIBA Publishing, London (2012)
RICS and Davis Langdon. *RICS Contracts in Use: A Survey of Building Contracts in Use During 2010*, RICS, London (2012)
Draft Supplementary Agreement for Consultant Switch or Novation (SA-SN-07)
RIBA Standard Agreement for the Appointment of an Architect (S-Con-07-A)
Supplementary Schedule for Contractor's Design Services (SupCD-07)

Note: RIBA Agreements are being updated and will be replaced by the RIBA Professional Services Contracts 2017

Cases

Alexander and another v *Mercouris* [1979] 1 WLR 1270	4.7
Alfred McAlpine Capital Projects Ltd v *Tilebox Ltd* [2005] BLR 271	5.52
Alfred McAlpine Homes North Ltd v *Property and Land Contractors Ltd* (1995) 76 BLR 59	7.33
Archivent Sales & Developments Ltd v *Strathclyde Regional Council* (1984) 27 BLR 98 (Court of Session, Outer House)	8.17
Balfour Beatty Building Ltd v *Chestermont Properties Ltd* (1993) 62 BLR 1	5.27
Bath and North East Somerset District Council v *Mowlem plc* [2004] BLR 153 (CA)	6.40
BFI Group of Companies Ltd v *DCB Integration Systems Ltd* [1987] CILL 348	5.52
Blyth and Blyth v *Carillion* (2001) 79 Con LR 142	1.18
Cavendish Square Holdings v *El Makdessi* and *ParkingEye Limited* v *Beavis*, Supreme Court 2015	5.52
CFW Architects v *Cowlin Construction Ltd* (2006) 105 Con LR 116 TCC	1.23
City Inn Ltd v *Shepherd Construction Ltd* [2008] CILL 2537 Outer House Court of Session	5.34

City of Westminster v *J Jarvis & Sons Ltd* (1970) 7 BLR 64 (HL)	5.44
Construction Partnership UK Ltd v *Leek Developments Ltd* [2006] CILL 2357 TCC	10.9
Co-operative Insurance Society v *Henry Boot Scotland and others* (2002) 84 Con LR 164	4.3
Croudace Ltd v *London Borough of Lambeth* (1986) 33 BLR 20 (CA)	7.37
Dawber Williamson Roofing Ltd v *Humberside County Council* (1979) 14 BLR 70	8.20
Department of Environment for Northern Ireland v *Farrans (Construction) Ltd* (1981) 19 BLR 1 (NI)	5.58
Ferson Contractors Ltd v *Levolux A T Ltd* [2003] BLR 118	10.23
F G Minter Ltd v *Welsh Health Technical Services Organisation* (1980) 13 BLR 1 (CA)	7.33
Greater London Council v *Cleveland Bridge and Engineering Co.* (1986) 34 BLR 50 (CA)	5.8
Greaves and Co. Contractors v *Baynham Meikle & Partners* (1975) 4 BLR 56	1.30
H Fairweather & Co. Ltd v *London Borough of Wandsworth* (1987) 39 BLR 106	5.33, 7.36
H W Nevill (Sunblest) Ltd v *William Press & Son Ltd* (1981) 20 BLR 78	5.44
Hall v *Van Der Heiden* [2010] EWHC 586 (TCC)	5.44
Henry Boot Construction (UK) Ltd v *Malmaison Hotel (Manchester) Ltd* (1999) 70 Con LR 32 (TCC)	5.34
Henry Boot Construction Ltd v *Alstom Combined Cycles* [2000] BLR 247	7.14
Holland Hannen & Cubitts (Northern) Ltd v *Welsh Health Technical Services Organisation* (1985) 35 BLR 1 (CA)	6.40
Impresa Castelli SpA v *Cola Holdings Ltd* (2002) CLJ 45	5.41
J F Finnegan Ltd v *Community Housing Association Ltd* (1995) 77 BLR 22 (CA)	5.55
Kruger Tissue (Industrial) Ltd v *Frank Galliers Ltd* (1998) 57 Con LR 1	9.10
Leedsford Ltd v *The Lord Mayor, Alderman and Citizens of the City of Bradford* (1956) 24 BLR 45 (CA)	4.16
London Borough of Barking & Dagenham v *Stamford Asphalt Co. Ltd* (1997) 82 BLR 25 (CA)	9.10
London Borough of Barking & Dagenham v *Terrapin Construction Ltd* [2000] BLR 479	4.15, 8.45
London Borough of Hounslow v *Twickenham Garden Developments* (1970) 7 BLR 81	10.14
London Borough of Merton v *Stanley Hugh Leach Ltd* (1985) 32 BLR 51 (ChD)	7.22
Melville Dundas Ltd v *George Wimpey UK Ltd* [2007] 1 WLR 1136 (HL)	10.23
Ministry of Defence v *Scott Wilson Kirkpatrick & Partners* [2000] BLR 20 (CA)	6.23
Multiplex Constructions (UK) Limited v *Honeywell Control Systems Limited (No. 2)* [2007] EWHC 447	5.25
National Trust for Places of Historic Interest and Natural Beauty v *Haden Young Ltd* (1994) 72 BLR 1 (CA)	9.10
Oksana Mul v *Hutton Construction Limited* [2014] EWHC 1797 (TCC)	6.41
Peak Construction (Liverpool) Ltd v *McKinney Foundations Ltd* (1970) 1 BLR 111 (CA)	5.18, 7.41
Pearce and High v *John P Baxter and Mrs A S Baxter* [1999] BLR 101 (CA)	6.59
Plant Construction v *Clive Adams Associates and JMH Construction Services* [2000] BLR 137 (CA)	4.3
Plymouth and South West Co-operative Society Ltd v *Architecture Structure and Management Ltd* [2006] CILL 2366 (TCC)	1.12
Reinwood Ltd v *L Brown & Sons Ltd* [2008] BLR 219 (CA)	5.57
Royal Brompton Hospital National Health Service Trust v *Frederick Alexander Hammond and others (No. 4)* [2000] BLR 75	1.23
Rupert Morgan Building Services (LLC) Ltd v *David Jervis and Harriet Jervis* [2004] BLR 18 (CA)	8.35
Skanska Construction (Regions) Ltd v *Anglo-Amsterdam Corporation Ltd* (2002) 84 Con LR 100	5.38, 5.41
Tameside Metropolitan Borough Council v *Barlow Securities Group Services Limited* [2001] BLR 113	3.2
Temloc Ltd v *Errill Properties Ltd* (1987) 39 BLR 30 (CA)	5.52
Trustees of Ampleforth Abbey Trust v *Turner & Townsend Project Management Limited* [2012] EWHC 2137	1.12
Viking Grain Storage Ltd v *T H White Installations Ltd* (1985) 33 BLR 103	1.29
Walter Lilly & Co. Ltd v *Giles Mackay & DMW Ltd* [2012] EWHC 649 (TCC)	5.34
Wates Construction (South) Ltd v *Bredero Fleet Ltd* (1993) 63 BLR 128	7.14
West Faulkner Associates v *London Borough of Newham* (1992) 61 BLR 81	10.14
Whittal Builders Co. Ltd v *Chester-le-Street District Council* (1987) 40 BLR 82	5.3

Legislation

Statutes

Arbitration Act 1996	11.20–11.21, 11.26
Companies Act 2006	3.4
Consumer Rights Act 2015	5.2
Contracts (Rights of Third Parties) Act 1999	3.26, 3.51–3.54
Defective Premises Act 1972	1.32, 4.7
Freedom of Information Act 2000	2.18
Housing Grants, Construction and Regeneration Act 1996	2.15, 2.20, 3.25, 8.1–8.2, 8.34, 8.37, 10.23, 11.8, 11.16–11.17
Law of Property Act 1925	3.48
Local Democracy, Economic Development and Construction Act 2009	2.15, 8.1, 11.8
Occupiers' Liability Acts 1957, 1984	5.45
Sale of Goods Act 1979	1.27, 8.17–8.18, 8.20
Sale and Supply of Goods Act 1994	1.27
Supply of Goods and Services Act 1982	1.28, 5.2
Unfair Contract Terms Act 1977	1.27

Statutory instruments

Construction (Design and Management) Regulations 2015	2.22, 3.18–3.20, 3.22, 4.28–4.32, 5.43, 6.5, 6.26, 6.35, 6.46, 10.12, 10.27
Health and Safety (Consultation with Employees) Regulations 1996	4.32
Joint Fire Code	9.28–9.29
Late Payment of Commercial Debts Regulations 2013	2.18
Public Contracts Regulations 2015	2.18, 10.7, 10.32
Unfair Terms in Consumer Contracts Regulations 1999	2.20

Clause Index *by paragraph*

DB16	Paragraph				
Recitals		2.1.3		4.20	
first	2.6	2.1.4		6.26	
second	2.8	2.2		4.11	
third	2.11, 3.31, 3.39–3.41, 4.4	2.2.1		4.12	
		2.2.2		6.37	
sixth	2.3	2.2.3		4.16, 6.11	
seventh	2.3	2.3		5.4, 5.8, 5.15	
		2.4		5.5	
Articles		2.5		5.39	
Article 1	4.1	2.5.1		5.4	
Article 2	2.6, 7.1, 7.3	2.5.2		5.39	
Article 3	3.24, 6.2	2.6		6.53, 6.55	
Article 5	4.28	2.7.1		3.43	
Article 6	4.28, 6.5	2.7.2		3.43	
Article 7	11.8	2.7.3		3.44, 6.2	
Article 8	11.2, 11.21, 11.29, 11.20, 11.36	2.7.4		3.45	
		2.7.5		3.45	
Article 9	11.20	2.8		2.9, 3.44, 6.13, 6.14	
		2.9		3.32, 6.7	
Section 1		2.10		3.38	
1.1	3.14, 3.17, 3.24, 3.59, 6.13, 6.14	2.10.1		3.32	
		2.11		4.3	
1.3	3.28	2.12		3.33	
1.4.6	3.17	2.12.2		3.33	
1.5	3.25	2.13		3.34	
1.6	3.26, 3.53–3.54	2.14		3.38	
1.7	3.25, 6.23	2.14.1		3.35	
1.7.1	3.25	2.14.2		3.36, 3.37	
1.7.2	3.25, 6.24	2.15.1		3.42, 4.23, 4.25	
1.7.3	6.24	2.15.2		4.23	
1.7.4	3.25, 10.9, 10.26	2.15.2.1		3.42, 4.26	
1.8.1	8.44	2.15.2.2		3.9, 4.26	
1.8.1.1	4.15	2.15.2.3		4.27	
1.8.1.2	4.45	2.16		4.19	
1.8.1.3	4.45	2.17.1		1.31, 2.10, 4.5–4.7	
1.8.2.1	11.29	2.17.2		1.32, 4.7	
1.8.2.2	4.46	2.17.3		4.7, 4.8	
1.8.2.3	4.46	2.18		4.22	
1.9	8.14, 8.32	2.21		8.19–8.20	
1.10	4.12, 6.7, 6.26, 6.32, 6.45	2.22		8.19	
		2.23.2		5.16	
1.11	3.27	2.24.1		5.21	
		2.24.2		5.22	
Section 2		2.24.3		5.22	
2.1	4.1, 4.3, 4.11	2.25		5.15, 5.20, 5.38	
2.1.1	4.19–4.20, 6.44	2.25.1		5.23	
2.1.2	4.19, 4.24	2.25.2		5.23	
		2.25.3		5.25	

2.25.4	5.26	3.5.1	6.26
2.25.5	5.28	3.6	6.29, 6.40
2.25.6.1	5.31, 5.32	3.7.1	6.23
2.25.6.2	5.31	3.7.2	6.23
2.25.6.3	5.26	3.7.3	6.23
2.26	5.24, 5.29	3.8	6.27–6.28
2.26.1	3.32, 3.35, 3.37,	3.9	4.27, 6.31
	4.26, 6.31, 6.36,	3.9.1	6.26, 6.31, 6.32
	10.34	3.9.2	6.26, 6.33
2.26.2	5.6, 6.38	3.9.3	6.31
2.26.2.1	3.37	3.9.4	6.26, 6.35
2.26.2.9	9.23, 9.30	3.10	5.6
2.26.3	3.63, 5.5	3.11	3.14
2.26.4	3.63	3.12	6.38
2.26.5	5.24, 8.37	3.13.1	6.39–6.40
2.26.6	3.35, 4.29, 5.3,	3.13.2	6.41, 6.44
	5.24, 6.7, 6.56	3.13.3	6.42–6.43
2.26.7	5.24	3.14	6.44
2.26.8	5.24	3.15	5.24
2.26.9	5.24	3.16	3.18, 4.29, 5.42,
2.26.11	5.24		10.11
2.26.12	5.24	3.16.1	4.29
2.26.13	4.25, 5.24	3.16.2	3.20, 4.28, 6.6,
2.27	3.20, 5.42–5.43,		6.9
	5.46, 6.10	3.16.3	3.19, 4.29
2.28	5.50, 5.56–5.57	3.16.4	4.31
2.29.1	5.54–5.55		
2.29.1.1	5.50, 5.53	**Section 4**	
2.29.1.2	5.53, 5.55	4.1	7.5
2.29.2	5.54–5.56	4.2	7.3, 7.39
2.29.2.2	5.55, 8.31	4.3	7.5, 8.25
2.29.3	5.57	4.6	2.2, 3.21, 8.5, 8.28
2.29.4	5.57	4.7.1	8.14
2.30	5.35	4.7.2	8.7, 8.39
2.31	5.36	4.7.3	8.7
2.32	5.36	4.7.4	8.7
2.33	5.37	4.7.5	8.7, 8.29
2.34	5.37	4.9	8.37
2.35	5.45, 6.57, 6.62	4.9.1	8.7, 8.29, 8.42
2.35.1	6.58	4.9.3	8.29
2.35.2	6.59	4.9.5	8.30
2.36	6.61	4.9.6	8.36
2.37	3.12, 5.42, 6.10	4.9.7	8.36
		4.10.1	8.30
Section 3		4.10.3	8.14, 8.39
3.1	6.2, 6.4	4.11	7.29, 8.37
3.2	6.3	4.11.1	8.37
3.3.1	4.17, 6.45	4.11.2	8.24, 8.37
3.4	3.22–3.23,	4.12	7.39, 8.8–8.12,
	6.46–6.47, 6.52		8.25
3.4.1	6.46	4.12.1	8.10
3.4.2.1	6.46, 8.20	4.12.1.1	8.9
3.4.2.3	6.46	4.12.1.2	8.9
3.4.2.4	6.46	4.12.1.3	8.9
3.4.2.5	3.63, 6.47	4.12.1.4	8.9, 8.16
3.4.3	6.47	4.12.2.1	8.11
3.5	6.26	4.12.2.2	8.11

Clause Index

4.12.2.3	8.11, 8.25	**Section 5**	
4.12.2.4	8.9, 8.11	5.1	4.26, 6.31
4.12.2.5	8.11	5.1.1	6.32
4.12.3	8.31	5.1.2	6.26, 6.33
4.12.3.1	8.12	5.2	6.53, 7.6
4.12.3.2	8.12	5.4.1	6.31
4.13	7.39, 8.8–8.12, 8.25	5.4.2	7.14, 7.15
		5.4.3	7.14
4.13.1	8.10	5.4.4	7.14
4.13.1.1	8.9	5.5	7.15
4.13.1.2	8.16	5.6	3.32, 3.37
4.13.1.3	8.9		
4.13.2.1	8.11	**Section 6**	
4.13.2.2	8.11	6.1	6.56, 9.2, 9.8
4.13.2.3	8.11, 8.25	6.2	6.56, 9.2, 9.9
4.13.2.4	8.11	6.3.1	9.10
4.13.2.5	8.11	6.3.2	9.10
4.13.3	8.31	6.3.3	9.10
4.13.3.1	8.12	6.3.4	9.9
4.13.3.2	8.12	6.4	9.3, 9.5
4.15	3.21, 8.22–8.23	6.4.1	9.5
4.15.1	8.22	6.4.1.2	9.7
4.15.2	8.22	6.5	9.4, 9.12–9.13
4.15.3	8.22	6.7	9.14
4.15.4	8.22	6.8	3.25, 9.15, 9.18, 9.25
4.15.5	8.22		
4.16.1	8.27	6.9.1	9.14
4.16.2	8.27	6.10	9.25
4.17	8.26	6.11.1	9.26
4.18	8.26	6.11.2	9.26
4.18.1	8.26	6.11.4	9.26
4.18.2	5.37, 5.45, 8.26	6.11.5	9.26
4.19	3.32, 7.19, 7.22, 7.29–7.30	6.12	9.5
		6.13.1	9.20
4.19.1	5.5, 7.19	6.13.2	9.21
4.20	7.18, 7.19	6.13.3	9.20
4.20.1	7.19	6.13.4	9.20
4.20.2	7.19	6.13.5.1	9.22
4.20.3	7.19	6.13.5.3	9.22
4.20.4	7.20	6.13.6	9.23
4.21	7.18, 7.29	6.14	9.24
4.21.1	3.32, 3.37, 4.26, 6.31, 6.36, 10.34	6.15	9.27
		6.15.1	4.10
4.21.2	3.35, 3.37, 5.6, 6.7	6.15.3	9.27
		6.17	9.28
4.21.2.2	6.38	6.18	9.28
4.21.4	4.25	6.19	9.28
4.21.5	3.35, 4.29, 5.3, 6.56	6.19.1	9.29
		6.19.2	9.29
4.23	7.17	6.20	9.28
4.24	5.54		
4.24.1	8.40	**Section 7**	
4.24.2	8.40	7.1	3.49, 6.8, 10.11
4.24.3	8.41	7.2	3.49
4.24.4	8.41	7.3	3.46
4.24.5	5.53, 6.23, 8.43	7.4	3.50
4.24.6	8.44	7A	3.26

7A.1	3.60	8.12	9.24
7B	3.26	8.12.1	10.33
7B.1	3.60	8.12.2	10.33
7C	3.58, 3.60	8.12.2.1	10.30
7D	3.58, 3.60	8.12.2.2	10.30
7E	3.62	8.12.2.3	10.30
7E.1	3.60	8.12.2.5	10.30
7E.1.2	3.60	8.12.3	10.33
Section 8		**Section 9**	
8.1	10.16	9.1	11.5
8.2.1	10.10	9.2	11.8–11.9
8.2.3	10.9, 10.26	9.2.2.1	11.12
8.3.1	10.7	9.3	11.21
8.4	10.6	9.4	11.21
8.4.1	10.7, 10.8	9.4.1	11.22
8.4.1.1	10.10–10.11	9.5	11.21, 11.27
8.4.1.2	10.11, 10.13	9.6	11.21
8.4.1.3	10.11	9.7	11.21
8.4.1.4	3.49, 6.45, 10.11	9.8	11.21
8.4.1.5	4.30, 10.11		
8.4.2	10.8	**Schedules**	
8.4.3	10.8	Schedule 1	2.9, 3.5, 3.44, 6.14,
8.5	10.7		6.15–6.20, 10.34
8.5.2	10.16	paragraph 1	6.15
8.5.3.1	10.16	paragraph 1.5	3.23
8.5.3.2	10.16	paragraph 2	6.16
8.5.3.3	10.16	paragraph 3	6.17
8.6	2.18, 10.7	paragraph 4	6.17
8.7.1	10.20, 10.22	paragraph 5	6.17
8.7.2.1	10.21	paragraph 7	6.19
8.7.2.2	10.21	paragraph 8.1	6.19
8.7.2.3	10.21	paragraph 8.2	6.20
8.7.3	10.23	paragraph 8.3	6.18
8.7.3.1	10.23		
8.7.3.2	10.23	Schedule 2	2.3, 2.16, 2.17, 3.3
8.7.4	10.24	Supplemental Provision 1	3.16, 3.23, 6.49,
8.7.5	10.24		10.34, 10.36
8.8.1	10.25	paragraph 1.1	6.49
8.9	10.6, 10.26–10.27	paragraph 1.1.1	6.52
8.9.1	10.26	paragraph 1.1.2	6.53
8.9.1.1	8.38	paragraph 1.2	6.53
8.9.1.2	3.49	paragraph 1.3.1	10.34
8.9.1.3	4.29	paragraph 1.3.2	10.34
8.9.2	6.34, 6.56, 10.26,	paragraph 1.3.3	10.34
	10.27	paragraph 1.4	6.54
8.9.4	10.26	paragraph 1.4.1	10.34
8.10	10.26	paragraph 1.4.2	10.34, 10.36
8.10.1	10.29	paragraph 1.4.3	10.35
8.11	9.24, 10.6	paragraph 1.5	6.52, 10.36
8.11.1	10.31, 10.32	Supplemental Provision 2	7.6, 7.7–7.11, 7.18,
8.11.1.2	3.32, 3.37, 4.26,		7.23
	10.31	paragraph 2.2	7.7
8.11.1.3	9.24	paragraph 2.3	5.16, 7.8
8.11.1.6	4.25	paragraph 2.5	7.10
8.11.2	10.31	Supplemental Provision 3	7.18, 7.24–7.28
8.11.3	2.18, 10.30, 10.32	paragraph 3.2	7.26

paragraph 3.4	7.25	Schedule 5	3.54
paragraph 3.6	7.28	Part 1	3.55
Supplemental Provision 4	5.17	paragraph 1.1	3.50
Supplemental Provision 5	4.2	paragraph 1.1.2	3.56
Supplemental Provision 6	4.32	paragraph 1.3	3.56
Supplemental Provision 7	7.4, 7.6, 7.12	paragraph 1.4	3.56
Supplemental Provision 8	7.6, 7.13	paragraph 5	3.56
Supplemental Provision 10	5.21, 11.1, 11.4	paragraph 6	3.56
Supplemental Provision 11	2.3, 2.18, 3.45	Part 2	3.55
Supplemental Provision 12	2.3	paragraph 1.1	3.57
		paragraph 1.2	3.57
Schedule 3	9.4	paragraph 10	3.57
Insurance Option A	5.37, 8.11, 9.15–9.17, 9.20, 9.25	Schedule 6	
A.1	9.22	Part 1	2.4, 3.21, 8.5, 8.28
Insurance Option B	5.37, 8.11, 9.15, 9.17, 9.20, 9.25	Part 2	2.4, 3.21
Insurance Option C	8.11, 9.10, 9.20, 9.25	Part 3	2.4, 3.21, 8.26
C.1	9.10, 9.18, 9.19	Schedule 7	7.39
C.2	5.37, 9.18	paragraph A.9	5.23, 7.41
		paragraph B.10	7.41
Schedule 4	6.43	paragraph C.6	7.41

Subject Index *by paragraph*

acceleration, 5.17, 8.10
access to site, 5.3–5.7, 6.4, 6.33
adjudication, 11.8–11.19, 11.29
advance payment bond, 2.4, 3.21, 8.5, 8.28
advance payments, 3.21, 8.5–8.6, 8.28
antiquities, 5.24
arbitration, 11.20–11.36
architect
   authority, 1.25
   consultant switch, 1.17–1.18
   as employer's agent, 1.14, 1.24, 1.25, 2.13
   liability, 1.16, 1.17, 1.18, 1.30
   novation, 1.16, 1.23
   provision of information, 1.23
   relations with contractor, 1.14–1.17, 1.25
   relations with employer, 1.14–1.15, 1.19–1.21, 1.24–1.25
   role, 1.14–1.15
   terms of engagement, 1.21–1.24
as-built information, 3.12, 5.43, 6.10
assignment of contractual rights, 3.47–3.49

bills of quantities, 3.8
bonds, 2.4, 3.21
   advance payment, 2.4, 3.21, 8.5, 8.28
   performance, 3.21, 3.46
   retention, 2.4, 3.21, 8.26
building information modelling (BIM), 3.17
Building Regulations, 4.20

CDM Regulations, 2.22, 3.18, 4.28–4.32, 10.12
changes, 6.30–6.36
   affecting completion date, 5.16
   affecting construction phase plan, 4.31
   to comply with planning requirements, 3.9
   contractor's obligations, 4.3
   to contractor's proposals, 4.26, 6.19–6.20
   discrepancy from requirements, 3.36–3.37
   to employer's requirements, 3.6, 3.33, 6.30–6.34
   omission of work, 5.26–5.27, 5.28, 7.14
   relevant events, 6.36
   site boundary, 3.32
   in statutory requirements, 4.26, 5.24
   valuation, 3.15, 5.16, 7.6–7.16, 7.23 (*see also* contract sum analysis)
charges and fees, 4.22
'chose in action', 3.48
client *see* employer
collaborative working, 2.17, 4.2
collateral warranties, 3.51, 3.58–3.63, 6.47
commencement date, 5.1–5.7
communication methods, 3.25, 6.23–6.24
completion contracts, 10.20, 10.22
completion date, 5.1–5.2, 5.12–5.15
   acceleration, 5.17
   affect of delays, 5.32
   affect of omission of work, 5.26–5.27, 5.28
   changes affecting, 5.16
   failure to meet, 5.50–5.51
confidential information, 3.45
consequential losses, 4.8
Construction (Design and Management) (CDM) Regulations, 2.22, 3.18, 4.28–4.32, 10.12
Construction Industry Council (CIC), 1.23, 3.17
construction phase plan, 3.19, 4.31, 6.5
Construction Supply Chain Payment Charter 2014, 2.18
consultant switch, 1.17–1.18
consumer contracts, 2.20
Consumer Rights Act 2015, 5.2
contract administration, 2.12, 6.1
contract documents, 3.1–3.63
   copies of, 3.44
   custody and control, 3.43–3.45
   discrepancies and errors, 3.31–3.38
   interpretations and definitions, 3.24–3.27
   priority, 3.28–3.30
   sub-contracts, 3.22–3.23

contract execution, 1.12, 3.2, 3.4
contract sum, 7.1
   adjustments, 2.14, 6.41, 7.2–7.5, 8.25, 8.31, 8.40
   fluctuations provision, 5.8, 5.51, 7.38–7.41, 8.10
   two-stage tendering, 1.10, 1.12
contract sum analysis, 2.14, 3.15, 7.14–7.16
contractor
   access to site, 5.3–5.7, 6.33
   'best endeavours' to prevent delay, 5.31
   compliance with employer's instructions, 6.26–6.28
   directly engaged by employer, 6.55–6.56
   duty to proceed 'regularly and diligently', 5.8, 10.13–10.14
   estimates for changes, 5.16, 7.7–7.11, 7.23
   information to be provided by, 6.9–6.11
   insolvency, 10.15–10.19
   interim payment applications, 7.24, 8.7, 8.30
   liability, 1.28–1.29, 1.31, 2.10, 3.56–3.57, 4.4–4.8
   losses and/or expenses, 7.17–7.37, 8.24
   notice of delay, 5.21–5.22
   obligations, 4.1–4.32
   relations with architect, 1.14–1.17, 1.25
   right of suspension, 8.37
   sequencing of work, 5.8
contractor's design documents, 3.12
   contractor's obligations, 4.3
   copies for employer, 3.44, 6.13
   discrepancies, 3.35
   submission procedure, 6.12–6.20
contractor's programme, 3.12, 5.9–5.11, 5.30
   (see also construction phase plan)
contractor's proposals, 2.8–2.9, 3.13–3.14
   changes, 4.26, 6.19–6.20
   custody and control of documents, 3.43–3.45
   discrepancies, 2.11, 3.35
   divergence from employer's requirements, 3.39–3.41
   divergence from statutory requirements, 3.42, 4.23–4.25
   form of submission, 3.11, 3.13
Contracts (Rights of Third Parties) Act 1999, 3.26, 3.51, 3.52–3.54
contractual rights, 3.47–3.49
cost-saving measures, 7.12
covenants, 3.10

custody of documents, 3.43

date for commencement, 5.1–5.7
date for completion see completion date
date of possession by the contractor, 5.1–5.7
daywork, valuation, 7.15–7.16
Defective Premises Act 1972, 1.32, 4.7
defective work, 6.38–6.44
   acceptance of, 6.41, 6.60, 8.31
   appearing after rectification period, 6.58–6.59, 6.62
   exclusion from interim payments, 8.14–8.15, 8.34–8.35
   making good, 6.57–6.63
   removal of, 6.39–6.40
   schedule, 5.48, 6.58–6.59
defects liability period see rectification period
definitions in documentation, 3.24–3.25
delays (see also extensions of time; relevant events)
   caused by employer, 5.13, 5.18–5.19, 5.24, 7.29
   concurrent events, 5.33–5.34
   mixed responsibility, 5.34
   notice of delay, 5.21–5.22
   permissions and approvals, 4.25, 5.24
   possession by contractor, 5.5–5.7
design
   contractor's see contractor's design documents
   employer's outline, 3.7, 3.33, 4.3
Design and Build Contract (DB16)
   changes from previous edition, 2.22–2.23
   comparison with SBC16, 2.21
   key features, 2.6–2.18
   when to use, 2.19
design defects, 1.29, 1.31, 4.4–4.5, 4.8
design liability, 1.26–1.32, 2.10
design submission procedure, 6.12–6.20
development permissions, 3.9, 4.20, 4.25, 5.24
discrepancies (documentation), 2.11, 3.31–3.38
dispute notice, 8.44–8.45
dispute resolution, 11.1–11.36
disruption of work, 7.30, 7.36 (see also suspension of work)
domestic clients, 2.20, 4.7
drawings see contractor's design documents

dwellings, 1.32, 2.20, 4.7

easements, 3.10
Emden formula, 7.35
employer
   acceptance of contractor's proposals, 4.4
   delays caused by, 5.13, 5.18–5.19, 5.24, 7.29
   indemnity, 9.2–9.3
   information to be provided by, 6.7–6.8
   insolvency, 10.29
   instructions, 6.21–6.29
   non-payment by, 8.36–8.38, 10.27–10.28
   obligations, 6.7–6.8
   outline design, 3.7, 3.33, 4.3
   payment notices, 8.29
   relations with architect, 1.14–1.15, 1.19–1.21, 1.24–1.25
   response to design documents, 6.16–6.17
   use of site during works, 5.4
employer's agent, 2.13, 6.2
   access to site, 6.4
   architect as, 1.14, 1.24, 1.25, 2.13
Employer's Liability (Compulsory Insurance) Act 1969, 9.6
employer's requirements, 2.8–2.9 (*see also* named sub-contractors)
   architect's role, 1.14
   changes, 3.6, 3.33, 6.30–6.34
   custody and control, 3.43–3.45
   discrepancies, 2.11, 3.14, 3.31–3.38
   divergence from statutory requirements, 4.23–4.25, 4.27
   documentation, 3.5–3.12
environmental performance measures, 7.13
estimates for changes *see* valuation of changes
extensions of time, 5.18–5.34
   assessment, 5.29–5.34
   following non-completion notice, 5.57
   for neutral events, 5.19
   procedure, 5.20–5.28
   resulting from changes, 5.16

fair payment provisions, 2.18
fees and charges, 4.22
final payment, 8.40–8.46
final statement, 8.40–8.46, 10.24

fire safety, 9.28–9.29
'fitness for purpose', 1.29–1.30, 4.6–4.7, 4.9
'fixed' price, 7.38
fluctuations provision, 5.8, 5.51, 7.38–7.41, 8.10
framework agreements, 1.9, 2.3
Freedom of Information Act 2000, 2.18

'guaranteed' price, 7.38
guarantees, 3.46

handover meeting, 5.49
health and safety documents, 3.18–3.20
health and safety file, 3.20, 5.43, 6.9
health and safety legislation, 4.28–4.32
Housing Grants, Construction and Regeneration Act (HGCRA) 1996, 2.15, 2.18, 8.29
   contractor's right of suspension, 8.37
   dispute resolution, 11.8, 11.16–11.17
   payment provisions, 8.1
Hudson formula, 7.35

indemnity, employer's, 9.2–9.3
industrial action, 5.24
insolvency
   contractor, 10.15–10.19
   employer, 10.29
inspections, 6.37, 6.38
instructions, employer's, 6.21–6.29
insurance, 9.1–9.31
   contractors engaged by employer, 6.56
   insurance of the works, 9.14–9.24
   partial possession, 5.37
   third parties, 9.1–9.31
interest on overdue payments, 8.36
interim payments, 8.3–8.4 (*see also* payments to the contractor)
   applications, 7.24, 8.7, 8.30
   ascertainment of amounts due, 8.8–8.15, 8.25–8.27
   due date, 8.29
interpretation of contract, 3.24–3.27

Joint Contracts Tribunal (JCT)
   collateral warranties, 2.4
   Design and Build Guide, DB/G, 2.5
   Design and Build Sub-Contract, 6.46

Joint Contracts Tribunal (JCT) *Continued*
   forms of contract, 1.7, 2.19
   Pre-Construction Services Agreement (General Contractor) (PCSA), 1.11
   SBC16, 2.19, 2.21
Joint Fire Code, 9.28–9.29

Late Payment of Commercial Debts (Interest) Act 1998, 8.36
Late Payment of Commercial Debts Regulations 2013, 2.18
LDEDCA *see* Local Democracy, Economic Development and Construction Act (LDEDCA) 2009
liquidated damages, 5.12, 5.52–5.58
   in case of partial possession, 5.37
   deduction from payments, 5.50–5.51, 5.56
   employer's notice to claim, 5.54–5.55
   repaid to contractor, 5.57–5.58
'listed items', 8.22–8.23
litigation, 11.30–11.36
local authorities, 2.22
   as employer, 2.2, 3.21
   termination of contractor's employment, 10.7, 10.32
   work in pursuance of statutory duties, 5.24
Local Democracy, Economic Development and Construction Act (LDEDCA) 2009
   dispute resolution, 11.8
   payment provisions, 2.15, 8.1, 8.14, 8.24
losses and expenses
   reimbursement, 7.17–7.37, 8.24
   relevant matters, 7.18, 7.29–7.30
   valuation, 7.23–7.28, 7.31–7.32

making good defects, 6.57–6.63
materials and goods
   'listed items', 8.22–8.23
   ownership, 6.46, 8.19
   payments to the contractor, 2.4, 3.21, 8.16–8.21
   price adjustment, 5.8, 5.51, 7.38–7.41, 8.10
   removal from site, 6.46, 8.19
   removal of defective work, 6.39–6.40
   samples, 6.11
   shortages, 5.24
   standards and quality, 1.28–1.29, 4.6–4.7, 4.11–4.16, 6.37
   substitution, 4.12–4.16
mediation, 11.5–11.7

named sub-contractors, 6.49–6.54
   contract documents, 3.23
   contractor's responsibility, 2.16
   delays caused by, 5.24
   itemised in contract sum analysis, 3.16
   termination of employment, 10.34–10.36
neutral events, 5.19, 5.24, 10.6
non-completion notice, 5.50, 5.57
non-standard terms, 3.29
novation, 1.16, 1.18, 1.23

occupation by employer, 5.35–5.41
omission of work, 5.26–5.27, 5.28, 7.14
opening up completed work, 6.38, 6.42
operation and maintenance manual, 6.10
outline design, 3.7, 3.33, 4.3

partial possession, 5.35–5.38
partnering provisions, 1.9, 2.17
pay less notice, 5.55, 8.30, 8.31, 8.35
payment applications, 7.24, 8.7
payment notices, 8.29
payments to the contractor, 2.14, 8.1 (*see also* contract sum analysis)
   ascertainment of amounts due, 8.8–8.15, 8.25–8.27
   deductions from amounts due, 8.14, 8.30–8.32
   employer's obligation to pay, 8.33–8.35
   fair payment provisions, 2.18
   final date for, 2.18
   final payment, 8.40–8.46
   following termination of employment, 10.23
   non-payment by employer, 8.36–8.38, 10.27–10.28
   partial possession, 5.37
   practical completion, 8.39–8.40
   procedure, 8.29–8.32
   retention sums, 5.37, 8.26–8.27
   withholding, 8.15, 8.30
performance bonds, 3.21, 3.46
performance indicators, 2.17

performance requirements, 3.7 (*see also* standards and quality)
phased completion *see* sectional completion
planning permission, 3.9, 4.20, 4.26, 5.24
possession by employer, 5.35–5.41
possession of the site by contractor, 5.1–5.7
postponement of work, 5.5–5.7
practical completion, 5.42–5.49
    information required, 3.12, 3.20, 5.43
    partial possession, 5.36
    payment on, 8.39–8.40
    procedure, 5.47–5.49
    use or occupation before, 5.39–5.41
pre-construction information, 3.19
price adjustment *see* fluctuations provision
principal contractor, 3.18, 4.29–4.30, 6.5, 6.9
principal designer, 3.18, 4.28–4.29, 6.6
priority of contract documents, 3.28–3.30
'privity of contract', 3.51
professional indemnity insurance (PII), 3.56, 4.10, 9.27
programme, contractor's, 3.12, 5.9–5.11, 5.30 (*see also* construction phase plan)
protocols, 3.17
provisional sums, 3.14, 4.22, 7.6
public bodies, 2.22–2.23
    as employer, 2.2, 3.21
    payment provisions, 2.18
    termination of contractor's employment, 10.7, 10.32
Public Contracts Regulations 2015, 2.18

quality and standards, 1.28–1.29, 4.6–4.7, 4.11–4.16, 6.37

'reasonable skill and care', 1.29
rectification period, 5.36, 5.45, 6.58, 6.62–6.63
relevant events, 5.24, 6.36
    after completion date, 5.27
    change in site boundary, 3.32
    notice by contractor, 5.22
relevant matters, 7.18, 7.24, 7.29–7.30
repudiation, 10.3, 10.5
restrictions on documentation, 3.45
retention bond, 2.4, 3.21, 8.26
retention of title clause, 8.16–8.17

retention sums, 5.37, 8.26–8.27
RIBA
    Contractor's Design Services Schedule, 1.22
    Standard Agreement 2010 (2012 revision): Architect, 1.22

Sale and Supply of Goods Act 1994, 1.27
Sale of Goods Act 1979, 1.27
samples, 6.11
SBC16, 2.19, 2.21
sectional completion, 5.1, 5.20, 5.51 (*see also* partial possession)
sequencing of work, 5.8
single-stage tendering, 1.13
site access, 5.3–5.7, 6.4, 6.33
site boundary, 3.32, 6.7
site constraints, 3.10
site investigation, 3.10
site manager, 6.3
site meeting minutes, 6.25
'snagging' lists, 5.48
special terms, 3.29
specialist contractors, 1.11
specifications, 3.7
stage payments, 8.3, 8.4, 8.9, 8.13, 8.15
standards and quality, 1.28–1.29, 4.6–4.7, 4.11–4.16, 6.37 (*see also* workmanship)
start date, 5.1–5.7
statutory requirements, 4.18–4.27
    changes in, 4.26, 5.24
    delays in obtaining permissions, 5.24
    divergence from, 3.42, 4.23–4.25
statutory undertaker's work, 5.24, 6.55
sub-contractors, 2.16 (*see also* named sub-contractors)
    contractor's responsibility for, 4.17
    documentation, 3.22–3.23
    employer's consent, 6.45
    form of contract, 6.46
    materials and goods, 8.18, 8.20
    third party rights and/or warranties, 3.62–3.63, 6.47
substitution, 4.12–4.16
supplemental provisions, 2.3, 2.17, 2.18, 3.3
    contractor's estimates, 7.7–7.11
    contractor's loss and/or expense, 7.23–7.28

supplemental provisions *Continued*
   contractor's obligations, 4.2
   named sub-contractors, 2.16
Supply of Goods and Services Act 1982, 1.28, 5.2
suspension of work (*see also* disruption of work)
   contractor's costs and expenses, 8.24
   delays in statutory approvals, 4.25
   due to changes in requirements, 6.34
   employer's failure to provide instructions, etc., 6.7
   grounds for termination, 10.10, 10.26, 10.31
   non-payment by employer, 8.37

tender process, 1.10–1.13
termination of contractor's employment, 10.4–10.25
   initiated by contractor, 10.26–10.30
   initiated by employer, 10.7–10.25
   named sub-contractors, 10.34–10.36
terms of engagement, architect's, 1.21–1.24
terrorism cover (insurance), 9.25–9.26
testing, 6.38, 6.42

third parties
   indemnity and insurance, 9.1–9.31
   rights, 3.26, 3.50–3.63, 6.47
'time is of the essence', 5.14
two-stage tendering, 1.10–1.12

Unfair Contract Terms Act 1977, 1.27
Unfair Terms in Consumer Contracts Regulations 1999, 2.20

valuation of changes, 3.15, 5.16, 7.6–7.16, 7.23
   (*see also* contract sum analysis)
valuation of design work, 3.16
valuation of loss and/or expense, 7.24–7.28, 7.31–7.32
valuation rules, 7.14–7.16
value-improvement measures, 7.12
VAT, 7.4

warranties, 3.51, 3.58–3.63, 6.47
weather, adverse, 5.24
working hours, 6.33
workmanship, 4.7, 4.11, 4.17, 6.11, 6.37